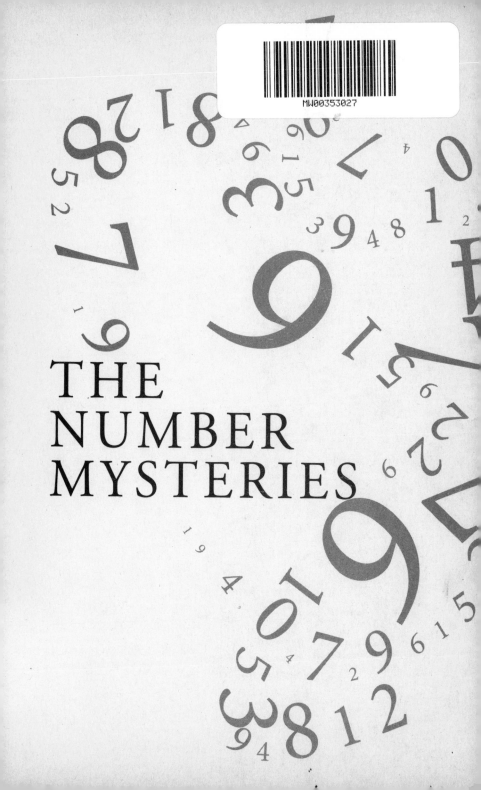

THE
NUMBER
MYSTERIES

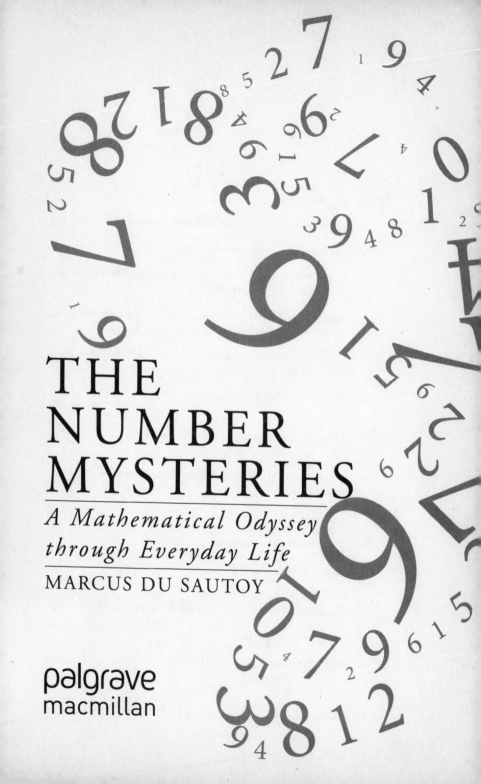

THE
NUMBER
MYSTERIES

A Mathematical Odyssey
through Everyday Life

MARCUS DU SAUTOY

palgrave
macmillan

THE NUMBER MYSTERIES
Copyright © Marcus du Sautoy, 2011.
All rights reserved.

First published in 2010 by Fourth Estate, an imprint of HarperCollins UK

First Published in the United States in 2011 by

PALGRAVE MACMILLAN®–a division of St. Martin's Press LLC, 175 Fifth Avenue, New York, NY 10010.

Palgrave Macmillan is the global academic imprint of the above companies and has companies and representatives throughout the world.

Palgrave® and Macmillan® are registered trademarks in the United States, the United Kingdom, Europe and other countries.

ISBN 978-0-230-11384-8

Library of Congress Cataloging-in-Publication Data

Du Sautoy, Marcus.
    The number mysteries : a mathematical odyssey through everyday life / Marcus du Sautoy.
        p.   cm.
    Includes index.
    ISBN 978-0-230-11384-8 (pbk.)
        1. Mathematics—Miscellanea.   I. Title.
QA99.D8   2011
510—dc22

2011000535

A catalogue record of the book is available from the British Library.

Design by Letra Libre, Inc.

First Palgrave Macmillan edition: May 2011

10 9 8 7 6 5 4 3 2 1

Printed in the United States of America.

CONTENTS

A NOTE ON WEBSITES

There are references throughout the book to external websites. All of these sites can be accessed in the usual way, by typing the address into a web browser. Alternatively, if you have a smartphone, you can use it to scan the QR codes printed by each website. QR codes are two-dimensional grids of black and white pixels and act like a bar code. On a QR code–compatible smartphone, you will need to download a QR reader first. To scan the code, launch the reader and hold the camera lens over the bar code in a sufficient amount of light. Most Android smartphones come with a built-in facility to scan QR codes. Those who own an iPhone can download a free QR reader from the App Store. Just enter the words *QR reader* into the search box. Blackberry phones running Blackberry Messenger 5.0 (or more recent versions) can also scan QR codes using the "Scan a Group Barcode" option in the Blackberry Messenger menu. A QR reader won't just be useful for enhancing your reading of the book. Originating in Japan, QR codes are being used increasingly across the globe on posters, as boarding passes for flights, and even on T-shirts to facilitate arranging a date with the person wearing the T-shirt.

INTRODUCTION

I s climate change a reality? Will the solar system suddenly fly apart? Is it safe to send your credit card number over the Internet? How can I beat the casino?

Ever since we've been able to communicate, we've been asking questions, trying to make predictions about what the future holds, and negotiating the environment around us. The most powerful tool that humans have created to navigate the wild and complex world we live in is mathematics.

From predicting the trajectory of a football to charting the population of lemmings, from cracking codes to winning at Monopoly, mathematics has provided the secret language to unlock nature's mysteries. But mathematicians don't have all the answers. There are many deep and fundamental questions we are still struggling to answer.

In each chapter of *The Number Mysteries,* I will take you on a journey through the big themes of mathematics, and at the end of each chapter, I reveal a mathematical mystery that no one has yet been able to solve. These are some of the great unsolved problems of all time.

But solving one of these conundrums won't just bring you mathematical fame—it will also bring you an astronomical fortune. An American businessman, Landon Clay, has offered a prize of a million dollars for the solution to each of these mathematical mysteries. You might think it strange that a businessman should want to hand out such big prizes for solving mathematical puzzles. But he knows that the whole of science, technology, the economy, and even the future of our planet relies on mathematics.

Each of the five chapters of this book introduces you to one of these million-dollar puzzles.

Chapter 1, "The Curious Incident of the Never-Ending Primes," takes as its theme the most basic object of mathematics: numbers. I will introduce you to the primes—the most important numbers in mathematics but also the most enigmatic. A mathematical million awaits the person who can unravel their secrets.

In chapter 2, "The Story of the Elusive Shape," we take a journey through the world's weird and wonderful shapes: from dice to bubbles, from tea bags to snowflakes. Ultimately, we tackle the biggest challenge of them all—what shape is our universe?

Chapter 3, "The Secret of the Winning Streak," will show you how the mathematics of logic and probability can give you the edge when it comes to playing games. Whether you like playing with Monopoly money or gambling with real cash, mathematics is often the secret to coming out on top. But there are some simple games that still fox even the greatest minds.

Cryptography is the subject of chapter 4, "The Case of the Uncrackable Code." Mathematics has often been the key to unscrambling secret messages. But I will reveal how you can use clever mathematics to create new codes that let you communicate securely across the Internet, send messages through space, and even read your friend's mind.

Chapter 5 is about what we would all love to be able to do: "The Quest to Predict the Future." I will explain how the equations of mathematics are the best fortune-tellers. They predict eclipses, explain why boomerangs come back, and ultimately tell us what the future holds for our planet. But some of these equations we still can't solve. The chapter ends with the problem of turbulence, which affects everything from David Beckham's free kicks to the flight of an airplane, yet is still one of mathematics' greatest mysteries.

The mathematics I present ranges from the easy to the difficult. The unsolved problems that conclude each chapter are so difficult that no one knows how to solve them. But I am a great believer in exposing people to the big ideas of mathematics. We get excited about literature when we encounter Shakespeare or Steinbeck. Music comes alive the first time we hear Mozart or Miles Davis. Playing Mozart yourself is tough, and Shakespeare can often be challenging, even for the experienced reader. But that doesn't mean that we should reserve the work of these great thinkers for the cognoscenti. Mathematics is just the same. So if some of the mathematics feels tough, enjoy what you can and remember the feeling of reading Shakespeare for the first time.

At school, we are taught that mathematics is fundamental to everything we do. In these five chapters, I want to bring mathematics to life to show you some of the great mathematics we have discovered to date. But I also want to give you the chance to test yourself against the biggest brains in history, as we look at some of the problems that remain unsolved. By the end, I hope you will understand that mathematics really is at the heart of all that we see and do.

One

THE CURIOUS INCIDENT OF THE NEVER-ENDING PRIMES

1, 2, 3, 4, 5 . . . it seems so simple: add 1, and you get the next number. Yet without numbers, we'd be lost. Arsenal vs. Manchester United— who won? We don't know. Each team scored lots of goals. Want to look something up in the index of this book? Well, the bit about winning the National Lottery is somewhere in the middle of the book. And the lottery itself? Hopeless without numbers. It's quite extraordinary how fundamental the language of numbers is to negotiating the world.

Even in the animal kingdom, numbers are fundamental. Packs of animals will base their decision to fight or flee on whether their group is outnumbered by a rival pack. Their survival instinct depends in part on a mathematical ability, yet behind the apparent simplicity of the list of numbers lies one of the biggest mysteries of mathematics.

2, 3, 5, 7, 11, 13 . . . These are the primes, the indivisible numbers that are the building blocks of all other numbers—the hydrogen and oxygen of the world of mathematics. These protagonists at the heart of the story of numbers are like jewels studded through the infinite expanse of numbers.

Yet despite their importance, prime numbers represent one of the most tantalizing puzzles we have come across in our pursuit of knowledge. Knowing how to find the primes is a total mystery because there seems to be no magic formula that gets you from one to the next. They are like buried treasure—and no one has the treasure map.

In this chapter, we will explore what we do understand about these special numbers. In the course of this journey, we will discover how different cultures have tried to record and survey primes and how musicians have exploited their syncopated rhythm. We will find out why the primes have been used to try to communicate with extraterrestrials and how they have helped to keep things secret on the Internet. At the end of the chapter, I unveil a mathematical enigma about prime numbers that will earn you a million dollars if you can crack it. But before we tackle one of the biggest conundrums of mathematics, let us begin with one of the great numerical mysteries of our time.

WHY DID BECKHAM CHOOSE THE NUMBER 23 SHIRT?

When David Beckham joined the Real Madrid soccer team in 2003, there was a lot of speculation about why he'd chosen to play in the number 23 shirt. It was a strange choice, many thought, since he'd been playing in the number 7 shirt for England and Manchester United. The trouble was that on the Real Madrid team, the number 7 shirt was already being worn by Raúl, and the Spaniard wasn't about to move over for this glamour-boy from England.

Many different theories were put forth to account for Beckham's choice, and the most popular was the Michael Jordan theory. Real Madrid wanted to break into the American market and sell lots of replica shirts to the huge US population. But soccer is not a popular game in the States. Americans like baseball and basketball—games that end with scores like 100 to 98 and in which there's invariably a winner. What's the point of a game that goes on for 90 minutes and can end with no side scoring or winning?

With this theory in mind, Real Madrid did its research and found that the most popular basketball player in the world was definitely Michael Jordan, the Chicago Bulls' most prolific scorer. Jordan sported the number 23 shirt for his entire career. All Real Madrid had to do was put 23 on the back of a soccer shirt, cross their fingers, and hope that the Jordan connection would work its magic and that they would break into the American market.

Others suggested a more sinister theory. Julius Caesar was assassinated by being stabbed in the back 23 times. Was Beckham's choice for his back a bad omen? Still others thought that maybe the choice was connected with Beckham's love of *Star Wars* (Princess Leia was imprisoned in Detention Block AA23 in the first *Star Wars* movie). Or was Beckham a secret member of the Discordianists, a modern cult that reveres chaos and has a cabalistic obsession with the number 23?

But as soon as I saw Beckham's number, a more mathematical solution immediately came to mind. 23 is a prime number. A prime number is a number that is divisible only by itself and 1. 17 and 23 are prime because they can't be written as two smaller numbers multiplied together, whereas 15 isn't prime because $15 = 3 \times 5$. Prime numbers are the most important numbers in mathematics because all other whole numbers are built by multiplying primes together.

Take 105, for example. This number is clearly divisible by 5. So I can write $105 = 5 \times 21$. 5 is a prime number—an indivisible number—but 21 isn't: I can write it as 3×7. So 105 can also be written as $3 \times 5 \times 7$. But this is as far as I can go: I'm down to the primes, the indivisible numbers from which the number 105 is built. I can do this with any number since every number is either prime and indivisible or not prime and divisible by smaller numbers multiplied together.

Figure 1.1

The primes are the building blocks of all numbers. Just as molecules are built from atoms, such as hydrogen and oxygen or sodium and chlorine, numbers are built from primes. In the world of mathematics, the numbers 2, 3, and 5 are like hydrogen, helium, and lithium. That's what makes them the most important numbers in mathematics. But they were clearly important to Real Madrid, too.

When I started looking a little closer at Real Madrid's soccer team, I began to suspect that perhaps they had a mathematician on the bench. A little analysis revealed that at the time of Beckham's move, all the Galácticos, the key players for Real Madrid, were playing in prime-number shirts: Carlos (the building block of the defense) wore number 3; Zidane (the heart of the midfield) was number 5; and Raúl and Ronaldo (the foundations of Real's strikers) sported numbers 7 and 11, respectively. So perhaps it was inevitable that Beckham also got a prime number, one that he has become very attached to. When he joined LA Galaxy, he insisted on taking his prime number with him in his attempt to woo the American public with the beautiful game.

This may sound totally irrational coming from a mathematician, someone who is meant to be a logical analytical thinker. However, I also play in a prime-number shirt for my soccer team, Recreativo Hackney, so I felt some connection with the man in 23. My Sunday League team isn't quite as big as Real Madrid and we didn't have a 23 shirt, so I chose 17—a rather nice prime, as we'll see later. But in our first season together, our team didn't do particularly well. We play in the London Super Sunday League Division 2, and that season we finished rock bottom. Fortunately, this is the lowest division in London, so the only way to go was up.

But how were we to improve our league standing? Maybe Real Madrid was on to something—was there some psychological advantage to be had from playing in a prime-number shirt? Perhaps too many of us were in nonprimes, like 8, 10, or 15. The next season, I persuaded the team to change our gear, and we all played in prime numbers: 2, 3, 5, 7 . . . all the way up to 43. It transformed us. We got promoted to Division 1, where we quickly learned that primes last only for one season. We were relegated back down to Division 2, and we are now on the lookout for a new mathematical theory to boost our chances.

A Prime-Number Fantasy Football Game

Download the PDF file for this game from the Number Mysteries website (http://www.fifthestate.co.uk/numbermysteries/). Each player cuts out three Subbuteo-style players and chooses different prime numbers to write on their backs. Use one of the Euclid soccer balls from chapter 2 (page 64).

The ball starts with a player from team 1. The aim is to make it past the three players on the opponent's team. The opponent chooses the first player to try to tackle team 1's player. Roll the die. The die has six sides: white 3, white 5, and white 7, and black 3, black 5, and black 7. The die will tell you to divide your prime and the prime of your opponent's player by 3, 5, or 7 and then take the remainder. If it is a white 3, 5, or 7, your remainder needs to match or beat the opposition. If it is black, you need to match or get less than your opponent.

To score, you must make it past all three players and then go up against a random choice of prime from the opposition. If at any point the opposition beats you, then possession switches

to the opposition. The person who has gained possession then
uses the player who won to try to make it through the opposi-
tion's three players. If team 1's shot at the goal is missed, then
team 2 takes the ball and gives it to one of its players.

The game can be played either against the clock or first
to three goals.

SHOULD REAL MADRID'S GOALKEEPER WEAR THE NUMBER 1 SHIRT?

If the key players for Real Madrid wear primes, then what shirt should the goalkeeper wear? Or, put mathematically, is 1 a prime? Well, yes and no. (This is just the sort of math question everyone loves—both answers are right.) Two hundred years ago, tables of prime numbers included 1 as the first prime. After all, it isn't divisible, since the only whole number that divides it is itself. But today, we say that 1 is not a prime because the most important thing about primes is that they are the building blocks of numbers. If I multiply a number by a prime, I get a new number. Although 1 is not divisible, if I multiply a number by 1, I get the number I started with, and on that basis, we exclude 1 from the list of primes and start at 2.

Clearly, Real Madrid wasn't the first to discover the potency of the primes. But which culture got there first—the ancient Greeks? The Chinese? The Egyptians? It turns out that mathematicians were beaten to the discovery of the primes by a strange little insect.

WHY DOES AN AMERICAN SPECIES OF CICADA LIKE THE PRIME 17?

In the forests of North America, there is a species of cicada with a very strange life cycle. For 17 years, these cicadas hide underground doing very little except sucking on the roots of the trees. Then in May of the seventeenth year, they emerge at the surface en masse to invade the forest—up to a million of them per acre.

The cicadas sing away to one another, trying to attract mates. Together, they make so much noise that local residents often move out for the duration of this invasion every 17 years. Bob Dylan was inspired to write his song "Day of the Locusts" when he heard the cacophony of cicadas that emerged in the forests around Princeton when he was collecting an honorary degree from the university in 1970.

After they've attracted a mate and become fertilized, the females each lay about six hundred eggs above ground. Then, after six weeks of partying, the cicadas all die and the forest goes quiet again for another 17 years. The next generation of eggs hatches in midsummer, and the nymphs drop to the forest floor before burrowing through the soil until they find a root to feed from. Then they wait another 17 years for the next great cicada party.

It's an absolutely extraordinary feat of biological engineering that these cicadas can count the passage of 17 years. It's very rare for any cicada to emerge a year early or a year too late. The annual cycle that most animals and plants work to is controlled by changing temperatures and the seasons. There is nothing that is obviously keeping track of the fact that the earth has gone around the sun 17 times and can then trigger the emergence of these cicadas.

For a mathematician, the most curious feature is the choice of number: 17, a prime number. Is it just a coincidence that these cicadas have chosen to spend a prime number of years hiding underground? It doesn't seem so. There are other species of cicada that stay underground for 13 years, and a few that prefer to stay there for 7 years—all prime numbers. Rather amazingly, if a 17-year cicada does appear too early, then it isn't out one year early, but generally four years early, apparently shifting to a 13-year cycle. There really does seem to be something about prime numbers that is helping these various species of cicada. But what is it?

While scientists aren't too sure, there is a mathematical theory that has emerged to explain the cicadas' addiction to primes. First, a few facts: A forest has, at most, one brood of cicadas, so the explanation isn't about sharing resources between different broods. In most years, a brood of prime-number cicadas emerges somewhere in the United States. However, 2009 and 2010 were cicada-free. In contrast, 2011 will see a massive brood of 13-year cicadas appearing in the southeastern United States. (Incidentally, 2011 is a prime, but I don't think the cicadas are that clever.)

The best theory to date for the cicadas' prime-number life cycle is the possible existence of a predator that also used to appear periodically in the forest, timing its arrival to coincide with the cicadas' and then feasting on the newly emerged insects. This is where natural selection kicks in, because cicadas that regulate their lives on a prime-number cycle are going to meet predators far less often than non-prime-number cicadas will.

For example, suppose that the predators appear every six years. Cicadas that appear every seven years will coincide with the predators only every 42 years. In contrast, cicadas that appear every eight years will coincide with the predators every 24 years; cicadas appearing every nine years will coincide even more frequently: every 18 years.

Figure 1.2   The interaction over a hundred years between populations of cicadas with a seven-year life cycle and predators with a six-year life cycle.

Figure 1.3   The interaction over a hundred years between populations of cicadas with a nine-year life cycle and predators with a six-year life cycle.

Across the forests of North America, there seems to have been real competition to find the biggest prime. The cicadas have been so successful that the predators have either starved or moved out, leaving the cicadas with their strange prime-number life cycle. But as we shall see, cicadas are not the only ones to have exploited the syncopated rhythm of the primes.

Cicadas vs. Predators

Download the PDF file for the cicada game from the Number Mysteries website (http://www.fifthestate.co.uk/numbermysteries/). Use the snakes-and-ladders board that can be downloaded from the same website. Cut out the predators and the two cicada families. Place the predators on the numbers in the 6 times table. Each player takes a family of cicadas. Take three standard six-sided dice. The roll of the dice will determine how often your family of cicadas appears. For example, if you roll an 8, then place cicadas on each number in the 8 times table. But if there is a predator already on a number, you can't place a cicada—for example, you can't place a cicada on 24 because it's already occupied by a predator. The winner is the person with the most cicadas on the board. You can vary the game by changing the period of the predator from 6 to some other number.

HOW ARE THE PRIMES 17 AND 29
THE KEY TO THE END OF TIME?

During the Second World War, the French composer Olivier Messiaen was incarcerated as a prisoner of war in Stalag VIII-A, where he discovered a clarinetist, a cellist, and a violinist among his fellow inmates. He decided to compose a quartet for these three musicians and himself on piano. The result was one of the great works of twentieth-century music: *Quatuor pour la fin du temps* (*Quartet for the End of Time*). It was first performed for inmates and prison officers inside Stalag VIII-A, with Messiaen playing a rickety upright piano they found in the camp.

In the first movement, called "Liturgie de Crystal," Messiaen wanted to create a sense of never-ending time, and the primes 17 and 29 turned out to be the key. While the violin and clarinet exchange themes representing birdsong, the cello and piano provide the rhythmic structure. In the piano part, there is a 17-note rhythmic sequence repeated over and over, and the chord sequence that is played on top of this rhythm consists of 29 chords. So as the 17-note rhythm starts for the second time, the chord sequence is just about two-thirds of the way through. The effect of the choice of prime numbers 17 and 29 is that the rhythmic and chordal sequences wouldn't repeat themselves until 17 × 29 notes through the piece.

It is this continually shifting music that creates the sense of timeless-ness that Messiaen was keen to establish—and he used the same trick as the cicadas with their predators. Think of the cicadas as the rhythm and the predators as the chords. The different primes, 17 and 29, keep the two out of sync so that the piece finishes before you ever hear the music repeat itself.

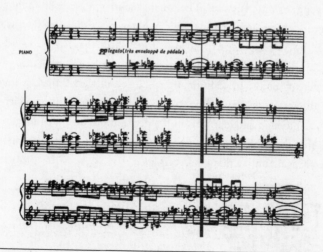

Figure 1.4   Messiaen's "Liturgie de Crystal" from the Quatuor pour la fin du temps. *The first vertical line indicates where the 17-note rhythm sequence ends. The second line indicates the end of the 29-note harmonic sequence. Property of Editions Durand, Paris. Reproduced by arrangement with G Ricordi & Co (London) Ltd, a division of Universal Music Publishing Group.*

Messiaen wasn't the only composer to have utilized prime numbers in music. Alban Berg also used a prime number as a signature in his music. Just like David Beckham, Berg sported the number 23—in fact, he was obsessed by it. For example, in his *Lyric Suite,* 23-bar sequences make up the structure of the piece. But embedded inside the piece is a representation of a love affair that Berg was having with a rich married woman. His lover was denoted by a ten-bar sequence that he entwined with his own signature 23, using the combination of mathematics and music to bring alive his affair.

Like Messiaen's use of primes in the *Quartet for the End of Time,* mathematics has recently been used to create a piece that although not timeless, nevertheless won't repeat itself for a thousand years. To mark the turn of the new millennium, Jem Finer, a founding member of The Pogues, decided to create a music installation in the East End of London that would repeat itself for the first time at the turn of the next millennium—3000. It's called, appropriately, *Longplayer.*

Finer started with a piece of music created with Tibetan singing bowls and gongs of different sizes. The original source music is 20 minutes and 20 seconds long, and by using some mathematics similar to the tricks employed by Messiaen, he expanded it into a piece that is a thousand years long. Six copies of the original source music are played simultaneously but at different speeds. In addition, every 20 seconds, each track is restarted a set distance from the original playback, but the amount by which each track is shifted is different. It is in the decision of how much to shift each track that the mathematics is used to guarantee that the tracks won't align perfectly again until a thousand years later.

You can listen to Longplayer *at http:// longplayer.org or by using your smartphone to scan this code.*

It's not just musicians who are obsessed with prime numbers: these numbers seem to strike a chord with practitioners in many different fields of the arts. The author Mark Haddon only used prime-number chapters in his best-selling book, *The Curious Incident of the Dog in the Night-Time*. The narrator of the story is a boy named Christopher, who has Asperger's syndrome. Christopher likes the mathematical world because he can understand how it will behave—the logic of this world means there are no surprises. Human interactions, though, are full of the uncertainties and illogical twists that Christopher can't cope with. As Christopher explains, "I like prime numbers . . . I think prime numbers are like life. They are very logical but you could never work out the rules, even if you spent all your lifetime thinking about them."

Prime numbers have even had an outing in the movies. In the futuristic thriller *Cube*, seven characters are trapped in a maze of rooms, which resembles a complex Rubik's Cube. Each room in the maze is cube-shaped with six doors leading to more rooms in the maze. The film begins with the characters waking up to find themselves inside this maze. They have no idea how they got there, but they have to find a way out. The problem is that some of the rooms are booby-trapped. The characters need to find some way of telling whether a room is safe before they enter it, for a whole array of horrific deaths await them, including being incinerated, getting covered in acid, and being cheese-wired into tiny cubes—as they discover when one of them is killed.

One of the characters, Joan, is a mathematical whiz, and she suddenly sees that the numbers at the entrance to each room hold the key to revealing whether a trap lies ahead. It seems that if any of the numbers at the entrance are prime, then the room contains a trap. "You beautiful brain," declares the leader of the group at this piece of mathematical deduction. It turns out that they also have to watch out for prime powers, but this proves to be beyond the clever Joan. Instead, they have to rely on one of their group who is an autistic savant, and he turns out to be the only one to make it out of the prime-number maze alive.

As the cicadas discovered, knowing your math is the key to survival in this world. Teachers who are having trouble motivating their mathematics class might find some of the gory deaths in *Cube* to be a great piece of propaganda for getting their students to learn their primes.

WHY DO SCIENCE FICTION WRITERS
LIKE PRIMES?

When science fiction writers want to get their aliens to communicate with Earth, they have a problem. Do they assume that their aliens are really clever and have picked up the local language, or that they've invented some clever Babel Fish–style translator that does the interpreting for them? Or do they just assume that everyone in the universe speaks English?

One solution that a number of authors have gone for is that mathematics is the only truly universal language, and the first words that anyone should speak in this language are its building blocks—the primes. In Carl Sagan's novel *Contact*, Ellie Arroway, who works for SETI (Search for Extraterrestrial Intelligence), picks up a signal that she realizes is not just background noise but a series of pulses. She guesses that they are binary representations of numbers. As she converts them into decimal numbers, she suddenly spots a pattern: 59, 61, 67, 71 . . . —all prime numbers. Sure enough, as the signal continues, it cycles through all the primes up to 907. This can't be random, she concludes. Someone is saying hello.

Many mathematicians believe that even if there is a different biology, a different chemistry, even a different physics on the other side of the universe, the mathematics will be the same. Anyone sitting on a planet orbiting Vega reading a math book about primes will still consider 59 and 61 to be prime numbers because, as the famous Cambridge mathematician G. H. Hardy put it, these numbers are prime "not because we think so, or because our minds are shaped in one way rather than another, but because it is so, because mathematical reality is built that way."

The primes may be numbers that are shared across the universe, but it is still interesting to wonder whether stories similar to those I've related are being told on other worlds. The way we have studied these numbers over the millennia has led us to discover important truths about them. And at each step on the way to discovering these truths, we can see the mark of a particular cultural perspective, the mathematical motifs of that period in history. Could other cultures across the universe have developed different perspectives, giving them access to theorems we have yet to discover?

Carl Sagan wasn't the first—and won't be the last—to suggest using the primes as a way of communicating. Prime numbers have even been used by NASA in its attempts to make contact with extraterrestrial intelligence. In 1974, the Arecibo radio telescope in Puerto Rico broadcast a message toward the globular star cluster M13, chosen for its huge number of stars so as to increase the chance that the message might fall on intelligent ears.

The message consisted of a series of 0s and 1s that could be arranged to form a black-and-white pixelated picture. The reconstructed image depicted the numbers from 1 to 10 in binary, a sketch of the structure of DNA, a representation of our solar system, and a picture of the Arecibo radio telescope itself. Considering that there were only 1,679 pixels, the picture is not very detailed. But the choice of 1,679 was deliberate because it contained the clue to setting up the pixels. 1,679 = 23 × 73, so there are only two ways to arrange the pixels in a rectangle to make up the picture. Arranging 23 rows of 73 columns produces a jumbled mess, but arrange them the other way—as 73 rows of 23 columns—and you get the result shown in Figure 1.5. The star cluster M13 is 25,000 light years away, so we're still waiting for a reply. Don't expect a response for another 50,000 years!

Figure 1.5   The message broadcast by the Arecibo radio telescope toward the star cluster M13.

Although the primes are universal, the way we write them has varied greatly throughout the history of mathematics, and is very culture-specific—as our whistle-stop tour of the planet will now illustrate.

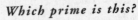

Which prime is this?

Figure 1.6

Some of the first mathematics in history was done in ancient Egypt, and this is how they wrote the number 200,201. As early as 6000 BC, people were abandoning nomadic life to settle along the Nile River. As Egyptian society became more sophisticated, the need grew for numbers to record taxes, measure land, and construct pyramids. Just as for their language, the Egyptians used hieroglyphs to write numbers. They had already developed a number system based on powers of 10, like the decimal system we use today. (The choice comes not from any special mathematical significance of the number, but from the anatomical fact that we have ten fingers.) But they had yet to invent the place-value system, which is a way of writing numbers so that the position of each digit corresponds to the power of 10 that the digit is counting. For example, the 2s in 222 all have different values according to their different positions. Instead, the Egyptians needed to create new symbols for each new power of 10:

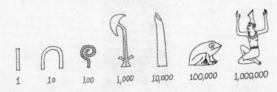

Figure 1.7   Ancient Egyptian symbols for powers of 10. 10 is a stylized heel bone, 100 a coil of rope, and 1,000 a lotus plant.

200,201 can be written quite economically in this way, but just try writing the prime 9,999,991 in hieroglyphs: you would need 55 symbols. Although the Egyptians did not realize the importance of the primes, they did develop some sophisticated math, including—not surprisingly—the formula for the volume of a pyramid and a concept of fractions. But their notation for numbers was not very sophisticated, unlike the one used by their neighbors, the Babylonians.

Which prime is this?

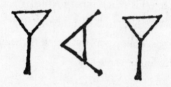

Figure 1.8

This is how the ancient Babylonians wrote the number 71. Like the Egyptian empire, the Babylonian empire was focused around a major river: the Euphrates. From 1800 BC, the Babylonians controlled much of modern Iraq, Iran, and Syria. To expand and run their empire, they became masters of managing and manipulating numbers. Records were kept on clay tablets, and scribes would use a wooden stick or stylus to make marks in the wet clay, which would then be dried. The tip of the stylus was wedge-shaped, or cuneiform—the name by which the Babylonian script is now known.

Around 2000 BC, the Babylonians became one of the first cultures to use the idea of a place-value number system. But instead of using powers of 10 like the Egyptians, the Babylonians developed a number system that worked in base 60. They had different combinations of symbols for all the numbers from 1 to 59. When they reached 60, they started a new "60s" column to the left and recorded one lot of 60, in the same way that in the decimal system, we place a 1 in the "10s" column when the units column passes 9. So the prime number shown in Figure 1.8 consists of one lot of 60 together with the symbol for 11, making 71. The symbols for the numbers up to 59 do have some hidden appeal to the decimal system because the numbers from 1 to 9 are represented by horizontal lines, but then 10 is represented by the following symbol:

Figure 1.9

The choice of base 60 is much more mathematically justified than the decimal system. It is a highly divisible number, which makes it very powerful for doing calculations. For example, if I have 60 beans, I can divide them up in a multitude of different ways:

$$60 = 30 \times 2 = 20 \times 3 = 15 \times 4 = 12 \times 5 = 10 \times 6$$

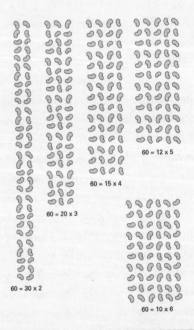

Figure 1.10   The different ways of dividing up 60 beans.

The Babylonians came close to discovering a very important number in mathematics: 0. If you wanted to write the prime number 3,607 in cuneiform, you had a problem. This is one lot of 3,600—or 60 squared—and 7 units, but if I write that down, it could easily look like one lot of 60 and 7 units—still a prime, but not the prime I want. To get around this, the Babylonians introduced a little symbol to denote that there were no 60s being counted in the 60s column. So 3,607 would be written as

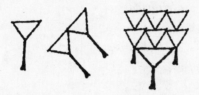

Figure 1.11

But they didn't think of 0 as a number in its own right. For them, it was just a symbol used in the place-value system to denote the absence of certain powers of 60. Mathematics would have to wait another 2,700 years until the seventh century AD, when the Indians introduced and investigated the properties of 0 as a number.

As well as developing a sophisticated way of writing numbers, the Babylonians are responsible for discovering the first method of solving quadratic equations—something every child is now taught at school. They also had the first inklings of Pythagoras's theorem about right-angled triangles. But there is no evidence that the Babylonians appreciated the beauty of prime numbers.

How to Count to 60 with Your Hands

We see many leftovers of the Babylonian base 60 today. There are 60 seconds in a minute, 60 minutes in an hour, 360 = 6 × 60 degrees in a circle. There is evidence that the Babylonians used their fingers to count to 60, in a quite sophisticated way.

Each finger is made up of three bones. Not counting the thumb, there are four fingers on each hand, so with the other

hand, you can point to any one of 12 different bones. The left hand is used to count to 12. The four fingers on the right hand are then used to keep track of how many lots of 12 you've counted. In total, you can count up to five lots of 12 (four lots of 12 on the right hand plus one lot of 12 counted on the left hand), so you can count up to 60.

For example, to indicate the prime number 29, you need to point to two lots of 12 on the right hand and then up to the fifth bone on the left hand.

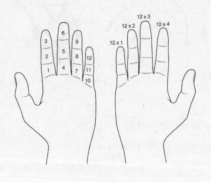

Figure 1.12

Which prime is this?

Figure 1.13

The Mesoamerican culture of the Maya was at its height from AD 200 to 900 and extended from southern Mexico through Guatemala to El Salvador. The

Maya had a sophisticated number system developed to facilitate the advanced astronomical calculations that they made, and this is how they would have written the number 17. In contrast to the Egyptians and Babylonians, the Maya worked with a base–20 system. They used a dot for 1, two dots for 2, three dots for 3, and four dots for 4. Just like a prisoner chalking off days on a prison wall, once they got to 5, instead of writing five dots, they would simply put a line through the four dots. A line, therefore, corresponds to 5.

It is interesting that the system works on the principle that our brains can quickly distinguish small quantities: we can tell the difference between one, two, three, and four things—but beyond that, it gets progressively harder. Once the Maya had counted to 19—three lines with four dots on top—they created a new column in which to count the number of 20s. The next column should have denoted the number of 400s (20 × 20), but bizarrely, it represents how many 360s (20 × 18) there are. This strange choice is connected with the cycles of the Mayan calendar. One cycle consists of 18 months of 20 days. (That's only 360 days. To make the year 365 days in length, they added an extra month of five "bad days," which were regarded as very unlucky.)

Interestingly, like the Babylonians, the Maya used a special symbol to denote the absence of certain powers of 20. Each place in their number system was associated with a god, and it was thought disrespectful to the god not to be given anything to hold, so a picture of a shell was used to denote nothing. The creation of this symbol to represent nothing was prompted by superstitious considerations as much as mathematical ones. Like the Babylonians, the Maya still did not consider 0 to be a number in its own right.

The Maya needed a number system to count very big numbers because their astronomical calculations spanned huge cycles of time. One cycle of time is measured by the so-called long count, which started on August 11, 3114 BC. It uses five placeholders and goes up to 20 × 20 × 20 × 18 × 20 days. That's a total of 7,890 years. Therefore, a significant date in the Mayan calendar will be December 21, 2012, when the Mayan date will turn to 13.0.0.0.0. Like kids in the back of a car waiting for the odometer to click over, Guatemalans are getting very excited by this forthcoming event—though some doom-mongers claim that it is the date of the end of the world.

Which prime is this?

Figure 1.14

Although these are letters rather than numbers, this is how to write the number 13 in Hebrew. In the Jewish tradition of gematria, the letters in the Hebrew alphabet all have a numerical value. Here, *gimel* is the third letter in the alphabet, and *yodh* is the tenth. So this combination of letters represents the number 13. The table opposite documents the numerical values of all the letters.

People who are versed in Kabbalah enjoy playing games with the numerical values of different words and seeing their interrelation. For example, my first name has the numerical value

mem		*resh*		*kaph*		*vav*		*samekh*		
40	+	200	+	20	+	6	+	60	=	326,

which has the same numerical value as "man of fame" . . . or, alternatively, "asses." One explanation for 666 being the number of the beast is that it corresponds to the numerical value of Nero, who was one of the most evil Roman emperors.

You can calculate the value of your name by adding up the numerical values in Table 1.1. To find other words that have the same numerical value as your name, visit http://billheidrick.com/works/hgemat .htm or use your smartphone to scan this code.

Hebrew letter	English equivalent	Numerical value
א, aleph	A	1
ב, beth	B	2
ג, gimel	G	3
ד, daleth	D	4
ה, he	H, E	5
ו, vav	V, U, O	6
ז, zayin	Z	7
ח, heth	Ch	8
ט, teth	T	9
י, yodh	I, Y, J	10
כ, kaph	K	20
ל, lamedh	L	30
מ, mem	M	40
נ, nun	N	50
ס, samek	S	60
ע, ayin	O, Ng	70
פ, pe	P	80
צ, sadhe	Tz	90
ק, qoph	Q	100
ר, resh	R	200
ש, sin	Sh	300
ת, tav	Th	400

Table 1.1

Although primes were not significant in Hebrew culture, related numbers were. Pick a number and look at all the numbers that divide into it (excluding the number itself) without leaving a remainder. If when you add up all these divisors you get the number you started with, then the number is called a perfect number. The first perfect number is 6. Apart from the number 6, the numbers that divide it are 1, 2, and 3. Add these together—1 + 2 + 3—and you get 6 again. The next perfect number is 28. The divisors of 28 are 1, 2, 4, 7, and 14, which add up to 28. According to the Jewish religion, the world was constructed in 6 days, and the lunar month used by the Jewish calendar was 28 days. This led to a belief in Jewish culture that perfect numbers had special significance.

The mathematical and religious properties of these perfect numbers were also picked up by Christian commentators. St. Augustine (354–430) wrote in his famous text *City of God* that "six is a number perfect in itself, and not because God created all things in six days; rather, the converse is true. God created all things in six days because the number is perfect."

Intriguingly, there are primes hidden behind these perfect numbers. Each perfect number corresponds to a special sort of prime number called a Mersenne prime (more on this later in the chapter). To date, we know of only 47 perfect numbers. The biggest has 25,956,377 digits. Perfect numbers that are even are always of the form $2^{n-1}(2^n - 1)$. And whenever $2^{n-1}(2^n - 1)$ is perfect, then $2^n - 1$ will be a prime number, and vice versa. We don't yet know whether there can be odd perfect numbers.

Which prime is this?

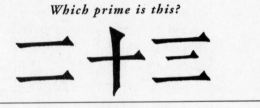

Figure 1.15

You might think that this is the prime number 5; it certainly looks like 2 + 3. However, the ＋ here is not a plus symbol—it is in fact the Chinese character for 10. The three characters together denote two lots of 10 and three units: 23.

This traditional Chinese form of writing numbers did not use a place-value system, but instead had symbols for the different powers of 10. An alternative system of representing numbers by bamboo sticks did use a place-value system and evolved from the abacus, on which when you reached ten, you would start a new column.

Here are the numbers from 1 to 9 in bamboo sticks:

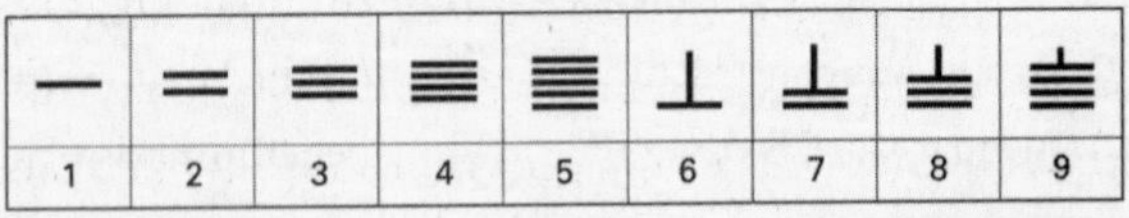

1	2	3	4	5	6	7	8	9

Figure 1.16

To avoid confusion, in every other column (namely the 10s, 1,000s, 100,000s, etc.) they turned the numbers around and laid the bamboo sticks vertically:

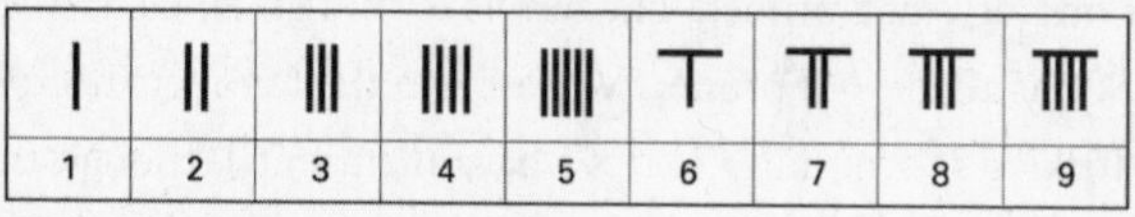

Figure 1.17

The ancient Chinese even had a concept of negative numbers, which they represented by different-colored bamboo sticks. The use of black and red ink in Western accounting is thought to have originated from the Chinese practice of using red and black sticks, although intriguingly, the Chinese used black sticks for negative numbers.

The Chinese were probably one of the first cultures to single out the primes as important numbers. They believed that each number had its own gender—even numbers were female, and odd numbers were male. They realized that some odd numbers were rather special. For example, if you have 15 stones, there is a way to arrange them into a nice-looking rectangle, in three rows of five. But if you have 17 stones, you can't make a neat array: all you can do is line them up in a straight line. For the Chinese, the primes were therefore the really macho numbers. The odd numbers that aren't prime, though they were male, were somehow rather effeminate.

This ancient Chinese perspective homed in on the essential property of being prime, because the number of stones in a pile is prime if there is no way to arrange them into a nice rectangle.

We've seen how the Egyptians used pictures of frogs to depict numbers; the Maya drew dots and dashes; the Babylonians made wedges in clay; the Chinese arranged sticks; and in Hebrew culture, letters of the alphabet stood for numbers. Although the Chinese were probably the first to single out the primes as important numbers, it was another culture that made the first inroads into uncovering the mysteries of these enigmatic numbers: the ancient Greeks.

HOW THE GREEKS USED SIEVES
TO COOK UP THE PRIMES

Here's a systematic way discovered by the ancient Greeks that is very effective at finding small primes. The task is to find an efficient method that will knock out all the nonprimes. Write down the numbers from 1 to 100. Start by striking out number 1. (As I have mentioned, though the Greeks believed 1 to be prime, in the twenty-first century, we no longer consider it to be.) Move to the next number, which is 2. This is the first prime. Now strike out every second number after 2. This effectively knocks out everything in the 2 times table, eliminating all the even numbers except for 2. Mathematicians like to joke that 2 is the odd prime because it's the only even prime . . . but perhaps humor isn't a mathematician's strong point.

Figure 1.18   Strike out every second number after 2.

Now take the lowest number that hasn't been struck out—in this case 3— and systematically knock out everything in the 3 times table.

Figure 1.19   Now strike out every third number after 3.

Because 4 has already been knocked out, we move next to the number 5 and strike out every fifth number on from 5. We keep repeating this process, going back to the lowest number *n* that hasn't yet been eliminated, and then striking out all the numbers *n* places ahead of it.

Figure 1.20   Finally, we are left with the primes from 1 to 100.

The beautiful thing about this process is that it is very mechanical—it doesn't require much thought to implement. For example, is 91 a prime? With this method, you don't have to think. 91 would have been struck out when you knocked out every seventh number on from 7 because 91 = 7 × 13. 91 often catches people off guard because we tend not to learn our 7 times table up to 13.

This systematic process is a good example of an algorithm, a method of solving a problem by applying a specified set of instructions—which is basically what a computer program is. This particular algorithm was discovered two millennia ago in one of the hotbeds of mathematical activity at the time: Alexandria, which is in present-day Egypt. Back then, Alexandria was one of the outposts of the great Greek empire and boasted one of the finest libraries in the world. It was during the third century BC that the librarian Eratosthenes came up with this early computer program for finding primes.

It is called the Sieve of Eratosthenes, because each time you knock out a group of nonprimes, it is as if you are using a sieve, setting the gaps between the wires of the sieve according to each new prime you move on to. First, you use a sieve where the wires are two apart—then three apart, five apart, and so on. The only problem is that the method soon becomes rather inefficient if you try to use it to find bigger and bigger primes.

As well as sieving for primes and looking after the hundreds of thousands of papyrus and vellum scrolls in the library, Eratosthenes also calculated the circumference of the earth and the distance of the earth to the sun and the moon. He calculated the sun to be 804 million stadia from the earth—although his unit of measurement perhaps makes judging the accuracy a little difficult. What size stadium are we supposed to use: Wembley or something smaller, like Loftus Road, the home ground of Fulham Soccer Club in London?

In addition to measuring the solar system, Eratosthenes charted the course of the Nile and gave the first correct explanation for why it kept flooding: heavy rains at the river's distant sources in Ethiopia. He even wrote poetry. Despite all this activity, his friends gave him the nickname Beta, because he never really excelled at anything. It is said that he starved himself to death after going blind in old age.

You can use your snakes-and-ladders board, downloadable from the Number Mysteries website, to put the Sieve of Eratosthenes into operation. Take a pile of pasta and place pieces on each of the numbers as you knock them out. The numbers left uncovered will be the primes.

HOW LONG WOULD IT TAKE TO WRITE A LIST OF ALL THE PRIMES?

Anyone who would try to write down a list of all the primes would be writing forever, because there are an infinite number of these numbers. What makes us so confident that we'll never come to the last prime, that there will always be another one waiting out there for us to add to the list? It is one of the greatest achievements of the human brain that with just a finite sequence of logical steps, we can capture infinity.

The first person to prove that the primes go on forever was a Greek mathematician living in Alexandria, named Euclid. He was a student of Plato's, and he also lived during the third century BC, though it appears he was about 50 years older than the librarian Eratosthenes.

To prove that there must be an infinite number of primes, Euclid started by asking whether, on the contrary, it was possible that there were in fact a finite number of primes. This finite list of primes would have to have the property that every other number could be produced by multiplying together primes from this finite list. For example, suppose that you thought that the list of all the primes consisted of just the three numbers 2, 3, and 5. Could every number be produced by multiplying together different combinations of 2s, 3s, and 5s? Euclid concocted a way to build a number that could never be captured by these three prime numbers. He began by multiplying together his list of primes to make 30. Then—and this was his act of genius—he added 1 to this number to make 31. None of the primes on his list—2, 3, or 5— would divide into it exactly. He always got a remainder of 1.

Euclid knew that all numbers are built by multiplying together primes, so what about 31? Since it can't be divided by 2, 3, or 5, there had to be some other primes not on his list that created 31. In fact, 31 is a prime itself, so Euclid had created a "new" prime. You might say that this new prime number could have just been added to the list. But Euclid could have then played the same trick again. However big the table of primes, Euclid could have just multiplied the list of primes together and added 1. Each time, he could have created a number that left a remainder of 1 upon division by any of the primes on the list, so this new number had to be divisible by primes not on the list. In this way, Euclid proved that no finite list could ever contain all the primes. Therefore, there must be an infinite number of primes.

Although Euclid could prove that the primes go on forever, there was one problem with his proof—it didn't tell you where the primes are. You might think that his method produces a way of generating new primes. After all, when we multiplied 2, 3, and 5 together and added 1, we got 31—a new prime. But it doesn't always work. For example, consider the list of primes 2, 3, 5, 7, 11, and 13. Multiply them all together: 30,030. Now add 1 to this number: 30,031. This number is not divisible by any of the primes from 2 to 13, because you always get a remainder of 1. However, it isn't a prime number since it is divisible by the two primes 59 and 509, and they weren't on our list. In fact, mathematicians still don't know whether the process of multiplying a finite list of primes together and adding 1 will infinitely often give you a new prime number.

There's a video available of my soccer team in their prime-number gear explaining why there are an infinite number of primes. Visit www.youtube.com/ watch?v=0LU4nkQKIN4 or use your smartphone to scan this code.

WHY ARE MY DAUGHTERS' MIDDLE NAMES 41 AND 43?

If we can't write down the primes in one big table, then perhaps we can try to find some pattern to help us to generate the primes. Is there some clever way to look at the primes you've got so far and know where the next one will be?

Here are the primes we discovered by using the Sieve of Eratosthenes on the numbers from 1 to 100: 2, 3, 5, 7, 11, 13, 17, 19, 23, 29, 31, 37, 41, 43, 47, 53, 59, 61, 67, 71, 73, 79, 83, 89, 97. The problem with the primes is that it can be really difficult to work out where the next one will be, because there don't seem to be any patterns in the sequence that will help us to locate them. In fact, they look more like a set of lottery ticket numbers than the building blocks of mathematics. Like waiting for a bus, you can have a huge gap with no primes and then suddenly several come along in quick succession. This behavior is very characteristic of random processes, as we shall see in Chapter 3.

Apart from 2 and 3, the closest that two prime numbers can be is two apart, like 17 and 19 or 41 and 43, since the number between each pair is always even and therefore not prime. These pairs of very-close primes are called twin primes. With my obsession for primes, my twin daughters almost ended up with the names 41 and 43. After all, if Chris Martin and Gwyneth Paltrow can call their baby Apple, and Frank Zappa can call his daughters Moon Unit and Diva Thin Muffin Pigeen, why can't my twins be 41 and 43? My wife was not so keen on the idea, however, so these have had to remain my "secret" middle names for the girls.

Although primes get rarer and rarer as you move out into the universe of numbers, it's extraordinary how often another pair of twin primes pops up. For example, after the prime 1,129, you don't find any primes in the next 21 numbers; then suddenly up pop the twin primes 1,151 and 1,153. And when you pass the prime 102,701, you have to plow through 59 nonprimes to get to the pair 102,761 and 102,763. The largest twin primes discovered by the beginning of 2009 have 58,711 digits. Given that it only takes a number with 80 digits to describe the number of atoms in the observable universe, these numbers are ridiculously large.

But are there more beyond these two twins? Thanks to Euclid's proof, we know that we're going to find an infinite number of primes. But are we going to keep coming across twin primes? As yet, nobody has come up with a clever proof like Euclid's to show whether there are an infinite number of these twin primes.

At one stage, it seemed that human twins might have been the key to unlocking the secret of prime numbers. In *The Man Who Mistook His Wife for a Hat,* Oliver Sacks describes the case of two real-life autistic savant twins who used the primes as a secret language. The twin brothers would sit in Sacks's clinic, swapping large numbers between themselves. At first, Sacks was mystified by their dialogue, but one night, he cracked the secret to their code. Studying up on some prime numbers of his own, he decided to test his theory. The next day, he joined the twins as they sat exchanging six-digit numbers. After a while, Sacks took advantage of a pause in the prime-number patter to announce a seven-digit prime, taking the twins by surprise. They sat thinking for a while, since this was stretching the limit of the primes they had been exchanging to date; then they smiled simultaneously, as if recognizing a friend.

During their time with Sacks, they managed to reach primes with nine digits. Of course, no one would find it remarkable if they were simply exchanging odd numbers or perhaps even square numbers, but the striking thing about what they were doing is that the primes are so randomly scattered. One explanation for how they managed it relates to another ability the twins had. Often, they would appear on television and impress audiences by identifying that, for example, October 23, 1901, was a Wednesday.

Working out the day of the week from a given date is done by something called modular, or clock, arithmetic. Maybe the twins discovered that clock arithmetic is also the key to a method that identifies whether a number is prime.

If you take a number, say 17, and calculate 2^{17}, then if the remainder when you divide this number by 17 is 2, that is good evidence that the number 17 is prime. This test for primality is often wrongly attributed to the Chinese; it was the seventeenth-century French mathematician Pierre de Fermat who proved that if the remainder isn't 2, then that certainly implies that 17 is not prime. In general, if you want to check that p is not a prime, then calculate 2^p and divide the result by p. If the remainder isn't 2, then p can't be prime. Some people have speculated that, given the twins' aptitude for identifying days of the week—which depends on a similar technique of looking at remainders upon division by 7—they may well have been using this test to find primes.

At first, mathematicians thought that if 2^p does have a remainder of 2 upon division by p, then p must be prime. But it turns out that this test does not guarantee primality. $341 = 31 \times 11$ is not prime, yet 2^{341} has a remainder of 2 upon division by 341. This example was not discovered until 1819, and it is possible that the twins might have been aware of a more sophisticated test that would wheedle out 341. Fermat showed that the test can be extended past powers of 2 by proving that if p is prime, then for any number n less than p, n^p always has a remainder of n when divided by the prime p. So if you find any number n for which this fails, you can throw out p as a prime impostor.

For example, 3^{341} doesn't have a remainder of 3 upon division by 341—it has a remainder of 168. The twins couldn't possibly have been checking through all numbers less than their candidate prime: there would be too many tests for them to run through. However, the great Hungarian prime-number wizard, Paul Erdös, estimated (though he couldn't prove it rigorously) that to test whether a number less than 10^{150} is prime, passing Fermat's test just once means that the chances of the number not being prime are as low as 1 in 10^{43}. So for the twins, probably one test was enough to give them the buzz of prime discovery.

PRIME-NUMBER HOPSCOTCH

This is a game for two players in which knowing your twin primes can give you an edge.

Write down the numbers from 1 to 100, or use the snakes-and-ladders board, which you can download from the Number Mysteries website. The first player takes a counter and places it on a prime number, which is, at most, five steps away from square 1. The second player takes the counter and moves it to a bigger prime that is, at most, five squares ahead of where the first player placed it. The first player follows suit, moving the counter to an even higher prime number that again is, at most, five squares ahead. The loser is the first player unable to move the counter according to the rules. The rules are the following: (1) the counter can't be moved farther than five squares ahead, (2) it must always be moved to a prime, and (3) it can't be moved backward or left where it is.

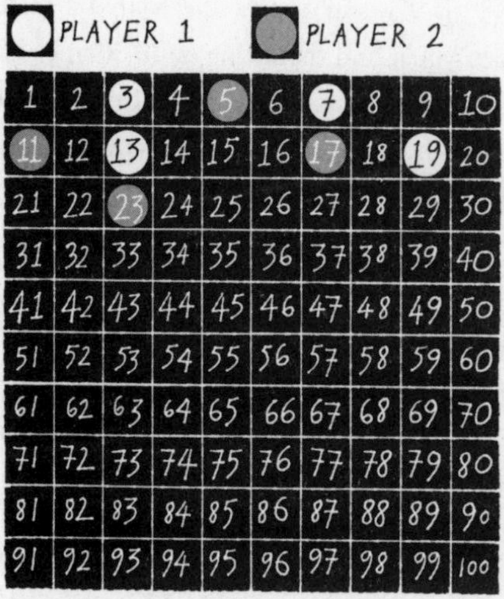

Figure 1.21   An example of a prime-number hopscotch game in which the maximum move is five steps.

Figure 1.21 shows a typical scenario. The first player has lost the game because the counter is at 23 and there are no primes in the five numbers ahead of 23. Could the first player have made a better opening move? If you look carefully, you'll see that once you've passed 5, there really aren't many choices. Whoever moves the counter to 5 is going to win because that player will, at a later turn, be able to move the counter from 19 to 23, leaving his or her opponent with no prime to move to. So the opening move is vital.

But what if we change the game a little? Let's say that you are allowed to move the counter to a prime that is, at most, seven steps ahead. Players can now jump a little farther. In particular, they can get past 23 because 29 is six steps ahead and within reach. Does your opening move matter this time? Where will the game end? If you play the game, you'll find that this time you have many more choices along the way, especially when there is a pair of twin primes.

At first glance, with so many choices, it looks like your first move is irrelevant. But look again. You lose if you find yourself on 89, because the next prime after 89 is 97—eight steps ahead. If you trace your way back through the primes, you'll find that being on 67 is crucial because here you get to choose which of the twin primes 71 and 73 you place the counter on. One is a winning choice; the other will make you lose the game because every move from that point on is forced on you. Whoever is on 67 can win the game, and it seems that 89 is not so important. So how can you make sure you get there?

If you continue tracing your way back through the game, you'll find that there's a crucial decision to be made for anyone on the prime 37. From there, you can reach my daughters' twin primes, 41 and 43. Move to 41, and you can guarantee winning the game. So now it looks as if the game is decided by whoever can get to the prime 37. Continuing to wind the game back in this way reveals that there is indeed a winning opening move. Put the counter on 5, and from there you can guarantee that you get all the crucial decisions that ensure you get to move the counter to 89 and win the game—because then your opponent can't move.

What if we continue to make the maximum permitted jumps even bigger: can we always be sure that the game will end? What if we allow each player to move a maximum of 99 steps—can we be sure that the game won't just go on forever because you can always jump to another prime within 99 of the last one? After all, we know that there are an infinite number of primes, so perhaps at some point, you can simply jump from one prime to the next.

It is actually possible to prove that the game does always end. However far you set the maximum jump, there will always be a stretch of numbers greater than the maximum jump containing no primes—it's there that the game will end. Let's look at how to find 99 consecutive numbers, none of which are prime. Take the number $100 \times 99 \times 98 \times 97 \times \ldots \times 3 \times 2 \times 1$. This number is known as 100 factorial and is written as 100! We're going to use an important fact about this number: if you take any number between 1 and 100, then 100! is divisible by this number. Look at this sequence of consecutive numbers:

$$100! + 2, 100! + 3, 100! + 4, \ldots, 100! + 98, 100! + 99, 100! + 100$$

100! + 2 is not prime because it is divisible by 2. Similarly, 100! + 3 is not prime because it is divisible by 3. (100! is divisible by 3, so if we add 3, it's still divisible by 3.) In fact, none of these numbers is prime. Take 100! + 53, which is not prime because 100! is divisible by 53, and if we add 53, the result is still divisible by 53. Here are 99 consecutive numbers, none of which is prime. The reason we started at 100! + 2 and not 100! + 1 is that with this simple method, we can deduce only that 100! + 1 is divisible by 1, and that won't help us to tell whether it's prime. (In fact, it isn't.)

So we know for certain that if we set the maximum jumps to 99, our prime-number hopscotch game will end at some point. But 100! is a ridiculously large number. The game would actually finish way before this point: the first place where a prime is followed by 99 nonprimes is 396,733.

Playing this game certainly reveals the erratic way in which the primes seem to be scattered through the universe of numbers. At first glance, there's no way of knowing where to find the next prime. But if we can't find a clever device for navigating from one prime to the next, can we at least come up with some clever formulas to produce primes?

 *This website has information about where the hopscotch game will end for larger and larger jumps: www.trnicely.net/ gaps/gaplist.html#MainTable. You can use your smartphone to scan this code.*

COULD RABBITS AND SUNFLOWERS BE USED TO FIND PRIMES?

Count the number of petals on a sunflower. Often, there are 89—a prime number. The number of pairs of rabbits after 11 generations is also 89. Have rabbits and flowers discovered some secret formula for finding primes? Not exactly. They like 89 not because it is prime, but because it is one of nature's other favorite numbers: the Fibonacci numbers. The Italian mathematician Fibonacci of Pisa discovered this important sequence of numbers in 1202 when he was trying to understand the way rabbits multiply (in the biological, rather than the mathematical, sense).

Fibonacci started by imagining a pair of baby rabbits—one male, one female. Call this starting point month 1. By month 2, these rabbits have matured into an adult pair, which can breed and produce in month 3 a new pair of baby rabbits. (For the purposes of this thought experiment, all litters consist of one male and one female.) In month 4, the first adult pair produce another pair of baby rabbits. Their first pair of baby rabbits has now reached adulthood, so there are now two pairs of adult rabbits and a pair of baby rabbits. In month 5, the two pairs of adult rabbits each produce a pair of baby rabbits. The baby rabbits from month 4 become adults. So by month 5, there are three pairs of adult rabbits and two pairs of baby rabbits, making five pairs of rabbits in total.

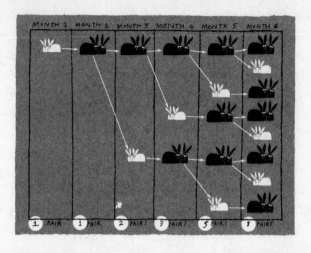

Figure 1.22   The Fibonacci numbers are the key to calculating the population growth of rabbits.

The number of pairs of rabbits in successive months is given by the following sequence: 1, 1, 2, 3, 5, 8, 13, 21, 34, 55, 89, . . . Keeping track of all these multiplying rabbits was quite a headache until Fibonacci spotted an easy way to work out the numbers. To get the next number in the sequence, you just add the two previous numbers. The bigger of the two is, of course, the number of pairs of rabbits up to that point. They all survive to the next month, and the smaller of the two is the number of adult pairs. These adult pairs each produce an extra pair of baby rabbits, so the number of rabbits in the next month is the sum of the numbers in the two previous generations.

Some readers might recognize this sequence from Dan Brown's novel *The Da Vinci Code.* They are in fact the first code that the hero has to crack on his way to the Holy Grail.

It isn't only rabbits and Dan Brown who like these numbers. The number of petals on a flower is often a Fibonacci number. Trilliums have three, pansies have five, delphiniums have eight, marigolds have 13, chicories have 21, pyrethrums have 34, and sunflowers often have 55 or even 89 petals. Some plants have flowers with twice a Fibonacci number of petals. These are plants, like some lilies, that are made up of two copies of a flower. And if your flower doesn't have a Fibonacci number of petals, then that's because a petal has fallen off . . . which is how mathematicians get around exceptions. (I don't want to be inundated with letters from irate gardeners, so I'll concede that there are a few exceptions that aren't just examples of wilting flowers. For example, the starflower often has seven petals. Biology is never as perfect as mathematics.)

As well as in flowers, you can find the Fibonacci numbers running up and down pine cones and pineapples. Slice across a banana, and you'll find that it's divided into three segments. Cut open an apple with a slice halfway between the stalk and the base, and you'll see a five-pointed star. Try the same with a Sharon fruit, and you'll get an eight-pointed star. Whether it's populations of rabbits or the structures of sunflowers or fruit, the Fibonacci numbers seem to crop up whenever there is growth happening.

The way shells evolve is also closely connected to these numbers. A baby snail starts off with a tiny shell, effectively a little one-by-one square house. As it outgrows its shell, it adds another room to the house and repeats the process as it continues to grow. Since it doesn't have much to go on, it simply adds a room whose dimensions are based on those of the two previous rooms, just as Fibonacci numbers are the sum of the previous two numbers. The result of this growth is a simple but beautiful spiral.

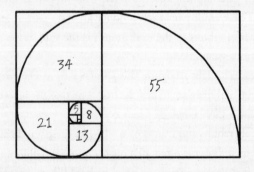

Figure 1.23   How to build a shell using Fibonacci numbers.

Actually, these numbers shouldn't be named after Fibonacci at all, because he was not the first to stumble across them. In fact, they weren't discovered by mathematicians at all, but by poets and musicians in medieval India. Indian poets and musicians were keen to explore all the possible rhythmic structures you can generate by using combinations of short and long rhythmic units. If a long sound is twice the length of a short sound, then how many different patterns are there with a set number of beats? For example, with eight beats you could do four long sounds or eight short ones. But there are lots of combinations between these two extremes.

In the eighth century AD, the Indian writer Virahanka took on the challenge of determining exactly how many different rhythms are possible. He discovered that as the number of beats goes up, the number of possible rhythmic patterns is given by the following sequence: 1, 2, 3, 5, 8, 13, 21, . . . He realized, just as Fibonacci did, that to get the next number in the sequence, you simply add together the two previous numbers. So if you want to know how many possible rhythms there are with eight beats, you go to the eighth number in the sequence, which is obtained by adding 13 and 21 to arrive at 34 different rhythmic patterns.

Perhaps it's easier to understand the mathematics behind these rhythms than to try to follow the increasing population of Fibonacci's rabbits. For example, to get the number of rhythms with eight beats, you take the rhythms with six beats and add a long sound or take the rhythms with seven beats and add a short sound.

There is an intriguing connection between the Fibonacci sequence and the protagonists of this chapter—the primes. Look at the first few Fibonacci numbers: 1, 1, 2, 3, 5, 8, 13, 21, 34, 55, 89, 144, . . . Every pth Fibonacci number—where p is a prime number—is itself prime. For example, 11 is prime, and the eleventh Fibonacci number is 89—a prime. If this always worked, it would be a great way to generate bigger and bigger primes. Unfortunately, it doesn't. The nineteenth Fibonacci number is 4,181, and although 19 is prime, 4,181 is not: it equals 37×113. No mathematician has yet proved whether an infinite number of Fibonacci numbers are prime numbers. This is another of the many unsolved prime-number mysteries in mathematics.

HOW CAN YOU USE RICE AND A CHESSBOARD TO FIND PRIMES?

Legend has it that chess was invented in India by a mathematician. The king was so grateful to the mathematician that he asked him to name any prize as a reward. The inventor thought for a minute, then asked for one grain of rice to be placed on the first square of the chessboard, two on the second, four on the third, eight on the fourth, and so on, so that each square got twice as many grains of rice as were on the previous square.

The king readily agreed, astonished that the mathematician wanted so little—but he was in for a shock. When he began to place the rice on the board, the first few grains could hardly be seen. But by the time he'd gotten to the sixteenth square, he already needed another kilogram of rice. By the twentieth square, his servants had to bring in a wheelbarrow full. He never reached the sixty-fourth and last square on the board. By that point, the total number of grains of rice would have been a staggering 18,446,744,073,709,551,615.

Figure 1.24   Repeated doubling makes numbers grow very quickly.

If we tried to repeat the feat at the heart of London, the pile of rice on the sixty-fourth square would stretch to a distance of 15 miles from the center and would be so high that it would cover all the buildings. In fact, there would be more rice in this pile than has been produced across the globe in the last millennium.

Not surprisingly, the king of India failed to give the mathematician the prize he had been promised and was forced into parting with half his fortune instead. That's one way math can make you rich.

But what has all this rice got to do with finding big prime numbers? Ever since the Greeks had proved that the primes go on forever, mathematicians had been on the lookout for clever formulas that might generate bigger and bigger primes. One of the best of these formulas was discovered by a French monk named Marin Mersenne. Mersenne was a close friend of Pierre de Fermat and René Descartes, and he functioned like a seventeenth-century Internet hub, receiving letters from scientists all across Europe and communicating ideas to those he thought could develop them further.

His correspondence with Fermat led to the discovery of a powerful formula for finding huge primes. The secret of this formula is hidden inside the story of the rice and the chessboard. As you count up the grains of rice from the first square of the chessboard, the cumulative total quite often turns out to be a prime number. For example, after three squares, there are $1 + 2 + 4 = 7$ grains of rice—a prime number. By the fifth square, there are $1 + 2 + 4 + 8 + 16 = 31$ grains of rice.

Mersenne wondered whether it would turn out to be true that whenever you landed on a prime-number square on the chessboard, the number of grains of rice up to that point might also be a prime. If it were, it would give you a way of generating bigger and bigger primes. Once you'd counted the prime-number grains of rice, you'd just move to this square on the chessboard and count the number of grains of rice up to this point, which Mersenne hoped would be an even bigger prime.

Unfortunately for Mersenne and for mathematics, his idea didn't quite work. When you look at the eleventh square of the chessboard, a prime-number square, then up to that point there are a total of 2,047 grains of rice. Sadly, 2,047 is not prime—it equals 23×89. But although Mersenne's idea didn't always work, it has led to some of the largest prime numbers that have been discovered.

THE GUINNESS BOOK OF PRIMES

During the reign of Queen Elizabeth I, the largest known prime number was the number of grains of rice on the chessboard up to and including the nineteenth square: 524,287. By the time Lord Nelson was fighting the Battle of Trafalgar, the record for the largest prime had gone up to the thirty-first square of the chessboard: 2,147,483,647. This ten-digit number was proved to be prime in 1772 by the Swiss mathematician Leonhard Euler, and it held the record until 1867.

By September 4, 2006, the record had gone up to the number of grains of rice that would be on the board up to square number 32,582,657, if we had a big enough chessboard. This new prime number has over 9.8 million digits, and it would take a month and a half to read it out loud. It was discovered not by some huge supercomputer but by an amateur mathematician using some software downloaded from the Internet.

The idea of this software is to utilize a computer's idle time to do computations. The program it uses implements a clever strategy that was developed to test whether Mersenne's numbers are prime. It still took a desktop computer several months to check Mersenne numbers with 9.8 million digits, but this is a lot faster than methods for testing whether a random number of this size is prime. By 2009, over ten thousand people had joined what has become known as the Great Internet Mersenne Prime Search, or GIMPS.

If you want your computer to join the GIMPS, you can download the software at www.mersenne.org, or scan this code with your smartphone.

Be warned, though, that the search is not without its dangers. One GIMPS recruit worked for a US telephone company and decided to enlist the help of 2,585 of the company's computers in his search for Mersenne primes.

The company began to get suspicious when its computers were taking five minutes rather than five seconds to retrieve telephone numbers. When the FBI eventually found the source of the slowdown, the employee owned up: "All that computational power was just too tempting for me," he admitted. The telephone company didn't feel sympathetic to the pursuit of science, and fired the employee.

After September 2006, mathematicians were holding their breath to see when the record would pass the ten million–digit barrier. The anticipation was not just for academic reasons—a prize of $100,000 was waiting for the person who got there first. The prize money was put up by the Electronic Frontier Foundation, a California-based organization that encourages collaboration and cooperation in cyberspace.

It was two more years before the record was broken. In a cruel twist of fate, two record-breaking primes were found within a few days of each other. The German amateur prime-number sleuth Hans-Michael Elvenich must have thought he'd hit the jackpot when his computer announced on September 6, 2008, that it had just found a new Mersenne prime with 11,185,272 digits. But when he submitted his discovery to the authorities, his excitement turned to despair—he had been beaten to it by 14 days. On August 23, Edson Smith's computer in the math department at the University of California at Los Angeles (UCLA) had discovered an even bigger prime—one with 12,978,189 digits. For UCLA, breaking prime-number records was nothing new. UCLA mathematician Raphael Robinson discovered five Mersenne primes in the 1950s, and two more were found by Alexander Hurwitz at the beginning of the 1960s.

The developers of the program used by GIMPS agreed that the prize money shouldn't simply go to the lucky person who was assigned that Mersenne number to check. Instead, $5,000 went to the developers of the software, $20,000 was shared among those who had broken records with the software since 1999, $25,000 was given to charity, and the rest went to Edson Smith in California.

If you still want to win money by looking for primes, don't worry about the fact that the ten million–digit mark has been passed. For each new Mersenne prime found, there is a prize of $3,000. But if it's the big money you're after, there is $150,000 on offer for passing one hundred million digits and $200,000 if you can make it to the billion-digit mark. Thanks to the ancient Greeks, we know that such record primes are waiting out there for someone to discover them. It's just a matter of how much inflation will have eaten into the worth of these prizes before someone eventually claims the next one.

How to Write a Number with 12,978,189 Digits

Edson Smith's prime is phenomenally large. It would take over three thousand pages of this book to record its digits, but luckily, a bit of mathematics can produce a formula that expresses the number in a much more succinct manner.

The total number of grains of rice up to the Nth square of the chessboard is

$$R = 1 + 2 + 4 + 8 + \ldots + 2^{N-2} + 2^{N-1}.$$

Here's a trick to find a formula for this number. It looks totally useless at first glance because it is so obvious: R = 2R − R. How on earth can such an obvious equation help in calculating R? In mathematics, it often helps to take a slightly different perspective, because then everything can suddenly look completely different.

Let's first calculate 2R. That just means doubling all the terms in the big sum. But the point is that if you double the number of grains of rice on one square, the result is the same as the number of grains on the next square along. So

$$2R = 2 + 4 + 8 + 16 + \ldots + 2^{N-1} + 2^{N}$$

The next move is to subtract R. This will just knock out all the terms of 2R except the last one:

$$R = 2R - R = (2 + 4 + 8 + 16 + \ldots + 2^{N-1} + 2^N) -$$
$$(1 + 2 + 4 + 8 + \ldots + 2^{N-2} + 2^{N-1})$$

$$= (2 + 4 + 8 + 16 + \ldots + 2^{N-1}) + 2^N - 1 - (2 + 4$$
$$+ 8 + \ldots + 2^{N-2} + 2^{N-1})$$

$$= 2^N - 1$$

So the total number of grains of rice up to the Nth square of the chessboard is $2^N - 1$, and this is the formula responsible for today's record-breaking primes. By doubling enough times and then taking 1 away from the answer, you might just hope to hit a Mersenne prime, as primes found using this formula are called.

HOW TO CROSS THE UNIVERSE WITH A DRAGON NOODLE

Rice is not the only food associated with exploiting the power of doubling to create large numbers. Dragon noodles, or la mian noodles, are traditionally made by stretching the dough between your arms and then folding it back again to double the length. Each time you stretch the dough, the noodle becomes longer and thinner, but you need to work quickly because the dough dries out quickly, disintegrating into a noodly mess.

Cooks across Asia have competed for the accolade of doubling the noodle length the most times, and in 2001, the Taiwanese cook Chang Hun-yu managed to double his dough 14 times in two minutes. The noodle he ended up with was so thin that it could be passed through the eye of a needle and would have stretched from Mr. Chang's restaurant in the center of Taipei to the outskirts of the city. When the noodle was cut, there were a total of 16,384 noodles.

This is the power of doubling, and it can very quickly lead to very big numbers. For example, if it were possible for Chang Hun-yu to have doubled his noodle 46 times, the noodle would have been the thickness of an atom and would have been long enough to reach from Taipei to the outer reaches of our solar system. Doubling the noodle 90 times would have gotten you from one side of the observable universe to the other. To get a sense of how big the current record prime number discovered in 2008 is, you would need to double the noodle 43,112,609 times and then take one noodle away.

WHAT ARE THE ODDS THAT YOUR TELEPHONE NUMBER IS PRIME?

One of the geeky things that mathematicians always do is check their telephone number to see whether it is prime. I moved homes recently and needed to change my telephone number. I hadn't had a prime telephone number at my previous house (house number 53, a prime), so I was hoping that at my new house (number 1, an ex-prime), I might be luckier.

The first number the phone company gave me looked promising, but when I put it into my computer and tested it, I found that it was divisible by 7. "I'm not sure I'm going to remember that number . . . any chance of another number?" I asked. The next number was also not prime—it was divisible by 3. (An easy test to see whether your number is divisible by 3 is this: add up all the digits of your telephone number, and if the number you get is divisible by 3, then so is the original number.) After about three more attempts, the exasperated telephone company employee snapped: "Sir, I'm afraid I'm just going to give you the next number that comes up." Alas, I now have an even telephone number, of all things!

So what were my chances of getting a prime telephone number? My number has eight digits. There is approximately a 1 in 17 chance that an eight-digit number is prime, but how does that probability change as the number of digits increases? For example, there are 25 primes under 100, which means that a number with two or fewer digits has a 1 in 4 chance of being prime. On average, as you count from 1 to 100, you get a prime every four numbers. But primes get rarer the higher you count.

This table shows the changes in probability:

Number of digits	Chance of getting a prime
1 or 2	1 in 4
3	1 in 6
4	1 in 8.1
5	1 in 10.4
6	1 in 12.7
7	1 in 15.0
8	1 in 17.4
9	1 in 19.7
10	1 in 22.0

Table 1.2

Though primes get rarer and rarer, they get rarer in a very regular way. Every time I add a digit, the probability decreases by about the same amount—2.3—each time. The first person to notice this was a 15-year-old boy. His name was Carl Friedrich Gauss (1777–1855), and he would go on to become one of the greatest names in mathematics.

Gauss made his discovery after being given a book of mathematical tables for his birthday. At the back of the book was a table of prime numbers. He was so obsessed with these numbers that he spent the rest of his life adding more and more figures to the tables in his spare time. Gauss was an experimental mathematician who liked to play around with data, and he believed that the way the primes thinned out would carry on in this uniform way however far you counted through the universe of numbers.

But how can you be sure that something strange won't suddenly happen when you hit one hundred–digit numbers, or one million–digit numbers? Would the probability still be the same as adding on 2.3 for each new digit, or could the probabilities suddenly start behaving totally differently? Gauss believed that the pattern would always be there, but it took until 1896 for him to be vindicated. Two mathematicians, Jacques Hadamard and Charles de la Vallée Poussin, independently proved what is now called the prime-number theorem: that the primes will always thin out in this uniform way.

Gauss's discovery has led to a very powerful model that helps to predict a lot about the behavior of prime numbers. It's as if, to choose the primes, nature used a set of prime-number dice with all sides blank except for one with a big *P* on it for prime.

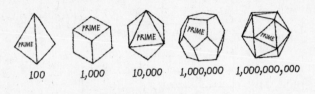

Figure 1.25   Nature's prime-number dice.

To decide whether each number is going to be prime, roll one of the dice. If it lands prime side up, then mark that number as prime; if it's blank side up, the number isn't prime. Of course, this is just a heuristic model—you can't make 100 indivisible just by the roll of dice. But it will give a set of numbers whose distribution is believed to be very like that of the primes. Gauss's prime-number theorem tells us how many sides there are on each die. So for three-digit numbers, use a six-sided die or a cube with one side prime. For four-digit numbers, use an eight-sided die—an octahedron. For five digits, a die with 10.4 sides . . . of course, these are theoretical dice because there isn't a polyhedron with 10.4 sides.

WHAT'S THE MILLION-DOLLAR PRIME PROBLEM?

The million-dollar question is about the nature of these dice: are the dice fair or not? Are the dice distributing the primes fairly through the universe of numbers, or are there regions where they are biased, sometimes giving too many primes, sometimes too few? The name of this problem is the Riemann hypothesis.

Bernhard Riemann was a student of Gauss's in the German city of Göttingen. He developed some very sophisticated mathematics that allow us to understand how these prime-number dice are distributing the primes. Using something called a zeta function, special numbers called imaginary numbers, and a fearsome amount of analysis, Riemann worked out the math that controls the fall of these dice. He believed from his analysis that the dice would be fair, but he couldn't prove it. To prove the Riemann hypothesis, that is what you have to do.

Another way to interpret the Riemann hypothesis is to compare the prime numbers to molecules of gas in a room. You may not know at any one instance where each molecule is, but the physics says that the molecules will be fairly evenly distributed around the room. There won't be a concentration of molecules in one corner and a complete vacuum in another. The Riemann hypothesis would have the same implication for the primes. It doesn't really help us to say where each particular prime can be found, but it does guarantee that they are distributed in a fair but random way through the universe of numbers. That kind of guarantee is often enough for mathematicians to be able to navigate the universe of numbers with a sufficient degree of confidence. However, until the million dollars is won, we'll never be quite certain what the primes are doing as we count our way further into the never-ending reaches of the mathematical cosmos.

Two

THE STORY OF THE ELUSIVE SHAPE

T he great seventeenth-century scientist Galileo Galilei once wrote, "The universe cannot be read until we have learnt the language and become familiar with the characters in which it is written. It is written in mathematical language, and the letters are triangles, circles and other geometrical figures, without which means it is humanly impossible to comprehend a single word. Without these, one is wandering about in a dark labyrinth."

This chapter presents the A to Z of nature's weird and wonderful shapes: from the six-pointed snowflake to the spiral of DNA, from the radial symmetry of a diamond to the complex shape of a leaf. Why are bubbles perfectly spherical? How does the body make such hugely complex shapes like the human lung? What shape is our universe? Math is at the heart of understanding how and why nature makes such a variety of shapes, and it also gives us the power to create new shapes, as well as the ability to say when there are no more shapes to be discovered.

It isn't only mathematicians who are interested in shapes: architects, engineers, scientists, and artists all want to understand how nature's shapes work. They all rely on the mathematics of geometry. The ancient Greek philosopher Plato put above his door a sign declaring, "Let no one ignorant of geometry enter here." In this chapter, I want to give you a passport to Plato's home, to the mathematical world of shapes. And at the end, I'll reveal another mathematical puzzle—one whose solution is worth another million dollars.

WHY ARE BUBBLES SPHERICAL?

Take a piece of wire and bend it into a square. Dip it in bubble mixture and blow. Why isn't it a cube-shaped bubble that comes out the other side? Or if the wire is triangular, why can't you blow a pyramid-shaped bubble? Why is it that, regardless of the shape of the frame, the bubble comes out as a perfect spherical ball? The answer is that nature is lazy, and the sphere is nature's easiest shape. The bubble tries to find the shape that uses the least amount of energy, and that energy is proportional to the surface area. The bubble contains a fixed volume of air, and that volume does not change if the shape changes. The sphere is the shape that has the smallest surface area that can contain that fixed amount of air. That makes it the shape that uses the least amount of energy.

Manufacturers have long been keen to copy nature's ability to make perfect spheres. If you're making ball bearings or shot for guns, getting perfect spheres could be a matter of life and death, since a slight imperfection in the spherical shape could lead to a gun backfiring or a machine breaking down. The breakthrough came in 1783 when an English plumber, William Watts, realized that he could exploit nature's predilection for spheres.

When molten iron is dropped from the top of a tall tower, like the bubble, the liquid droplets form into perfect spheres during their descent. Watts wondered whether, if you stuck a vat of water at the bottom of the tower, you could freeze the spherical shapes as the droplets of iron hit the water. He decided to try his idea out in his own house in Bristol. The problem was that he needed the drop to be higher than three floors to give the falling molten iron time to form into spherical droplets.

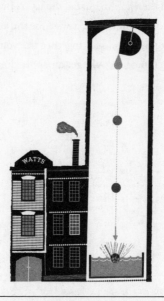

Figure 2.1 William Watts's clever use of nature to make spherical ball bearings.

So Watts added another three stories on top of his house and cut holes in all the floors to allow the iron to fall through the building. The neighbors were a bit shocked by the sudden appearance of this tower on the top of his home, despite his attempts to give it a Gothic twist with the addition of some castle-like trim around the top. But so successful were Watts's experiments that similar towers soon shot up across England and America. His own shot tower stayed in operation until 1968.

Although nature uses the sphere so often, how can we be sure that there isn't some other strange shape that might be even more efficient than the sphere? It was the great Greek mathematician Archimedes who first proposed that the sphere was indeed the shape with the smallest surface area containing a fixed volume. To try to prove this, Archimedes began by producing formulas for calculating the surface area of a sphere and the volume enclosed by it.

Calculating the volume of a curved shape was a significant challenge, but he applied a cunning trick: he sliced the sphere with parallel cuts into many thin layers, and then approximated the layers by disks. Now, he knew the formula for the volume of a disk: it was just the area of the circle times the thickness of the disk. By adding together the volumes of all these different-sized disks, Archimedes could get an approximation for the volume of the sphere.

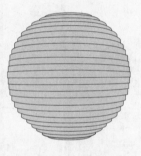

Figure 2.2   A sphere can be approximated by stacking different-sized disks on top of one another.

Then came the clever bit. If he made the disks thinner and thinner until they were infinitesimally thin, the formula would give an exact calculation of the volume. It was one of the first times that the idea of infinity was used in mathematics, and a similar technique would eventually become the basis for the mathematics of the calculus developed by Isaac Newton and Gottfried Leibniz nearly two thousand years later.

Archimedes went on to use this method to calculate the volumes of many different shapes. He was especially proud of the discovery that if you put a spherical ball inside a cylindrical tube of the same height, then the volume of the air in the tube is precisely half the volume of the ball. He was so excited by this that he insisted a cylinder and a sphere should be carved on his gravestone.

Although Archimedes had successfully found a method to calculate the volume and surface area of the sphere, he didn't have the skills to prove his hunch that it is the most efficient shape in nature. Amazingly, it was not until 1884 that the mathematics became sophisticated enough for the German Hermann Schwarz to prove that there is no mysterious shape with less energy that could trump the sphere.

HOW TO MAKE THE WORLD'S ROUNDEST SOCCER BALL

Many sports are played with spherical balls: tennis, cricket, snooker, soccer. Although nature is very good at making spheres, humans find it particularly tricky. This is because most of the time, we make the balls by cutting shapes from flat sheets of material that then have to be either molded or sewn together. In some sports, a virtue is made of the fact that it's hard to make spheres. A cricket ball consists of four molded pieces of leather sewn together, and so it isn't truly spherical. The seam can be exploited by a bowler to create unpredictable behavior as the ball bounces off the pitch.

In contrast, table-tennis players require balls that are perfectly spherical. The balls are made by fusing together two celluloid hemispheres, but the method is not very successful since over 95 percent are discarded. Ping-Pong ball manufacturers have great fun sorting the spheres from the misshapen balls. A gun fires balls through the air, and any that aren't spheres will swing to the left or to the right. Only those that are truly spherical fly dead straight and get collected on the other side of the firing range.

How, then, can we make the perfect sphere? In the buildup to the soccer World Cup in 2006 in Germany, there were claims by manufacturers that they had made the world's most spherical soccer ball. Soccer balls are very often constructed by sewing together flat pieces of leather, and many of the soccer balls that have been made over the generations are assembled from shapes that have been played with since ancient times. To find out how to make the most symmetrical soccer ball, we can start by exploring "balls" built from a number of copies of a single symmetrical piece of leather, arranged so that the assembled solid shape is symmetrical. To make it as symmetrical as possible, the same number of faces should meet at each point of the shape. These are the shapes that Plato explored in his *Timaeus*, written in 360 BC.

What are the different possibilities for Plato's soccer balls? The one requiring fewest components is made by sewing together four equilateral triangles to make a triangular-based pyramid called a tetrahedron—but this doesn't make a very good soccer ball because there are so few faces. As we shall see in chapter 3, this shape may not have made it onto the soccer-ball pitch, but it does feature in other games that were played in the ancient world.

Another configuration is the cube, which is made of six square faces. At first glance, this shape looks rather too stable for a soccer ball, but actually, its structure underlies many of the early soccer balls. The very first World Cup soccer ball used in 1930 consisted of 12 rectangular strips of leather grouped in six pairs and arranged as if assembling a cube. Although now rather shrunken and unsymmetrical, one of these balls is on display at the National Museum of Football in Preston, in the north of England. Another rather extraordinary soccer ball that was also used in the 1930s is also based on the cube and has six H-shaped pieces cleverly interconnected.

Figure 2.3   Some early designs for soccer balls.

Let's go back to equilateral triangles. Eight of them can be arranged symmetrically to make an octahedron, effectively by fusing two square-based pyramids together. Once they are fused together, you can't tell where the join is.

The more faces there are, the rounder Plato's soccer balls are likely to be. The next shape in line after the octahedron is the dodecahedron, made from 12 pentagonal faces. There is an association here with the 12 months of the year, and ancient examples of these shapes have been discovered with calendars carved on their faces. But of all Plato's shapes, it's the icosahedron, made out of 20 equilateral triangles, that approximates best to a spherical soccer ball.

Plato believed that together, four of these five shapes were so fundamental that they were related to the four classical elements, the building blocks of nature: the tetrahedron, the spikiest of the shapes, was the shape of fire; the stable cube was associated with earth; the octahedron was air; and the roundest of the shapes, the icosahedron, was slippery water. The fifth shape, the dodecahedron, Plato decided represented the shape of the universe.

Figure 2.4   The Platonic solids were associated with the building blocks of nature.

How can we be sure that there isn't a sixth soccer ball Plato might have missed? It was another Greek mathematician, Euclid, who in the climax to one of the greatest mathematical books ever written, proved that it's impossible to sew together any other combinations of a single symmetrical shape to make a sixth soccer ball to add to Plato's list. Called simply *The Elements,* Euclid's book is probably responsible for founding the analytical art of logical proof in mathematics. The power of mathematics is that it can provide 100 percent certainty about the world, and Euclid's proof tells us that, as far as these shapes go, we have seen everything—there really are no other surprises waiting out there that we've missed.

 *You can visit the Number Mysteries website and download PDF files that contain instructions for making each of Plato's five soccer balls. Make a goal out of card stock and see how good the different shapes are for finger soccer.*

    Try some of the tricks in this video: http://video.yahoo .com/watch/15164/554045. You can also access this video by using your smartphone to scan this code.

HOW ARCHIMEDES IMPROVED ON PLATO'S SOCCER BALLS

What if you tried to smooth out some of the corners of Plato's five soccer balls? If you took the 20-faced icosahedron and chopped off all the corners, then you might hope to get a rounder soccer ball. In the icosahedron, five triangles meet at each point, and if you chop off the corners, you get pentagons. The triangles with their three corners cut off become hexagons, and this so-called truncated icosahedron is in fact the shape that has been used for soccer balls ever since it was first introduced in the 1970 World Cup finals in Mexico. But are there other shapes made from a variety of symmetrical patches that could make an even better soccer ball for the next World Cup?

It was in the third century BC that the Greek mathematician Archimedes set out to improve on Plato's shapes. He started by looking at what happens if you use two or more different building blocks as the faces of your shape. The shapes still needed to fit neatly together, so the edges of each type of face had to be the same length. That way, you'd get an exact match along the edge. He also wanted as much symmetry as possible, so all the vertices—the corners where the faces meet—had to look identical. If two triangles and two squares met at one corner of the shape, then this had to happen at every corner.

The world of geometry was forever on Archimedes's mind. Even when his servants dragged a reluctant Archimedes from his mathematics to the baths to wash himself, he would spend his time drawing geometrical shapes in the embers of the chimney or in the oils on his naked body with his finger. Plutarch described how "the delight he had in the study of geometry took him so far from himself that it brought him into a state of ecstasy."

It was during these geometric trances that Archimedes came up with a complete classification of the best shapes for soccer balls, finding 13 different ways that such shapes could be put together. The manuscript in which Archimedes recorded his shapes has not survived, and it is only from the writings of Pappus of Alexandria, who lived some five hundred years later, that we have any record of the discovery of these 13 shapes. They nonetheless go by the name of the Archimedean solids.

Some he created by cutting bits off the Platonic solids, like the classic soccer ball. For example, if you snip the four ends off a tetrahedron, the original triangular faces then turn into hexagons, while the faces revealed by the cuts are four new triangles. So four hexagons and four triangles can be put together to make something called a truncated tetrahedron:

Figure 2.5

In fact, 7 of the 13 Archimedean solids can be created by cutting bits off Platonic solids, including the classic soccer ball of pentagons and hexagons. More remarkable was Archimedes's discovery of some of the other shapes. For example, it is possible to put together 30 squares, 20 hexagons, and 12 ten-sided figures to make a symmetrical shape called a great rhombicosidodecahedron:

Figure 2.6

It was one of these 13 Archimedean solids that was behind the new Zeitgeist ball introduced at the World Cup in Germany in 2006 and heralded as the world's roundest soccer ball. Made up of 14 curved pieces, the ball is actually structured around the truncated octahedron. If you take the octahedron made up of eight equilateral triangles and cut off the six vertices, the eight triangles become hexagons, and the six vertices are replaced by squares:

Figure 2.7

Perhaps future World Cups might feature one of the more exotic of Archimedes's soccer balls. My choice would be the snub dodecahedron, made up of 92 symmetrical pieces—12 pentagons and 80 equilateral triangles:

Figure 2.8

Even to the end, Archimedes's mind was on things mathematical. In 212 BC, the Romans invaded his hometown of Syracuse. He was so engrossed in drawing diagrams to solve a mathematical conundrum that he was completely unaware of the fall of the city around him. When a Roman soldier burst into his home with sword brandished, Archimedes pleaded to at least be able to finish his calculations before he ran him through. "How can I leave this work in such an imperfect state?" he cried. But the soldier was not prepared to wait for the QED, and hacked Archimedes down in midtheorem.

Pictures of all 13 Archimedean solids can be found at http://mathworld.wolfram .com/ArchimedeanSolid.html or by using your smartphone to scan this code.

WHAT SHAPE DO YOU LIKE YOUR TEA?

Shapes have become a hot issue not just for soccer-ball manufacturers but also for the tea drinkers of England. For generations, the British were content with the simple square, but now teacups are swimming with circles, spheres, and even pyramid-shaped tea bags in the nation's drive to brew the ultimate cup.

The tea bag was invented by mistake at the beginning of the twentieth century by a New York tea merchant, Thomas Sullivan. He'd sent customers samples of tea in small silken bags, but rather than removing the tea from the bags, customers assumed they were meant to put the whole bag in the water. It took until the 1950s for the British to be convinced to take on such a radical change to their tea-drinking habits, but today, it is estimated that over one hundred million tea bags are dunked each day in the United Kingdom.

For years, the trusty square had allowed tea drinkers to make a cup of tea without the hassle of having to wash out used tea leaves from teapots. The square is a very efficient shape—it was easy to make square tea bags, and there was no wastage of unused bits of bag material. For 50 years, PG Tips, the leading manufacturer of tea bags, stamped out billions of tea bags a year in its factories up and down the United Kingdom.

But in 1989, its main rival, Tetley, made a bold move to capture the market by changing the shape of the tea bag: Tetley introduced circular bags. Although the change was little more than an aesthetic gimmick, it worked. Sales of the new shape soared. PG Tips realized that it had to go one better if it was to retain its customers. The circle might have excited patrons, but it was still a flat, two-dimensional figure. So the team at PG decided to take a leap into the third dimension.

The PG Tips team knew that the British are an impatient lot when it comes to their tea. On average, the bag stays in the cup for just 20 seconds before being hoisted out. If you cut open the average two-dimensional bag after it has been dunked for just 20 seconds, you'll find that the tea in the middle is completely dry, not having had time to get in contact with the water. The researchers at PG believed that a three-dimensional bag would behave like a mini teapot, giving all the leaves the chance to make contact with the water. They even enlisted a thermofluids expert from the University of London's Imperial College to run computer models to confirm their belief in the power of the third dimension to improve the flavor of tea.

Then came the next step in development: what shape? A selection of different three-dimensional shapes were prepared for consumer testing. They experimented with cylindrical tea bags and ones that looked like Chinese lanterns, as well as perfect spheres. The sphere is quite attractive because, as the bubble knows, it's the three-dimensional shape that, for a given enclosed volume, requires the minimum amount of material to make the bag. But it's also an extremely difficult shape to manufacture, especially if you are starting with a flat sheet of muslin—as anyone who has tried to wrap a soccer ball at Christmas will testify.

Starting with a flat piece of paper, three-dimensional shapes with flat faces were the obvious things to consider, and PG Tips began by looking at the shapes that Plato and Archimedes had described over two thousand years ago. As sports manufacturers had discovered, a football made out of pentagons and hexagons approximates a sphere very well, but it was the shape at the other end of the spectrum that started to interest the tea-bag developers. The four-sided tetrahedron, or triangular-based pyramid, actually encloses the least volume for a given surface area. On the positive side, it is the shape that requires the fewest number of faces to make (there is no way to put together three flat faces to make a three-dimensional shape).

PG Tips was obviously keen not to have too much bag material wasted, so the shape had to be efficient, as well as visually attractive. On top of that, because PG Tips was trying to cater to a nation that drinks over one hundred million cups of tea a day, the bag had to be a shape that could be knocked out at a fast rate: the company didn't want factories full of workers sewing together four little triangles to make pyramids. The breakthrough came when someone came up with an extremely beautiful and elegant way to make a pyramid-shaped tea bag.

Consider how a potato-chip bag is made. A cylindrical tube is sealed at the bottom, filled with chips, and then sealed in the same direction along the top. But look what happens if instead of sealing the top in the same direction as the seal at the bottom, you twist the bag 90 degrees and then seal it. Suddenly, you're holding a tetrahedral bag in your hand. The tetrahedron has six edges: two where the seals have been made, and four that link the two seals—an edge runs from the end of each seal to each end of the opposite seal.

It's a beautifully efficient way to make a pyramid. Replace the chip bag with a tea bag that you seal with this twist, and you've got pyramidal tea bags. There is no wastage of material, and a machine can churn out these bags at a rate of two thousand a minute—fast enough to meet the United Kingdom's tea-drinking demands. The machine was so innovative that it made it onto a list of the top one hundred patents filed in the twentieth century.

After four years of development, the pyramid tea bag was launched in 1996. Not only did it turn out to be efficient, but consumers thought the shape had a modern, funky feel to it. The new advertising certainly made a welcome change to the troop of dressed-up monkeys that the company had relied on for years to sell its tea, and PG Tips regained its top spot in the tea-bag sales charts. But if tetrahedrons have brought out the taste of tea, then another of the Platonic solids is the shape of something rather more sinister.

WHY MIGHT CATCHING AN ICOSAHEDRON KILL YOU?

In 1918, the Spanish flu pandemic killed at least fifty million people—far more than the casualties of the First World War. Such devastation concentrated scientists' minds on determining the mechanism of this dangerous disease, and they soon realized that the cause was not bacteria but something much smaller that couldn't be seen under the microscopes of the time. They called these new agents "viruses," after the Latin word for poison.

Uncovering the true nature of these viruses had to wait for the development of a new technology, called X-ray diffraction, which gave scientists a way of penetrating the underlying molecular structure of the organisms that were causing such havoc. A molecule can be visualized as a collection of Ping-Pong balls joined together with toothpicks. Although this sounds too simple to be real science, every chemistry lab is equipped with ball-and-stick kits to help students and researchers explore the structure of the molecular world. In X-ray diffraction, a beam of X-rays is passed through the material being investigated, and the X-rays are deflected through various angles by the molecules they encounter. The pictures that are produced are a bit like the shadow you get when you shine a light on one of these ball-and-stick structures.

Math was a powerful ally in the battle to unravel the information contained in these shadows. The game is to identify what three-dimensional shapes could possibly give rise to the two-dimensional shadows that X-ray diffraction produced. Quite often, progress depends on finding the best angle at which to "shine the light" and reveal the molecule's true character. A silhouette of someone's head from the front gives little information beyond whether that person's ears stick out, but a profile tells you much more about who you're looking at. It's the same with molecules.

Having cracked the structure of DNA, Francis Crick and James Watson, along with Donald Caspar and Aaron Klug, turned their attention to what the two-dimensional pictures from X-ray diffraction could reveal about viruses. To their surprise, they found shapes full of symmetry. The first images showed dots arranged in triangles, which implied that the virus had a three-dimensional shape that could be spun by a third of a turn and look the same. When the biologists looked in the mathematicians' cabinet of shapes, it was the Platonic solids that seemed the best candidates for the form of these viruses.

The problem was that all five of Plato's shapes had an axis about which you could spin the shape by a third of a turn so that all the faces realign. It was only when the biologists obtained another diffraction image that they got a view that enabled them to pin down the shapes of these viruses more precisely. Suddenly, dots arranged in pentagons appeared, and that allowed them to home in on one of the more interesting of Plato's dice: the icosahedron—the shape made of 20 triangles with five triangles meeting at each point.

Viruses like symmetrical shapes because symmetry provides a very simple means for them to multiply, and that is what makes viral diseases so infectious—in fact, that's what *virulent* means. Traditionally, symmetry has been something people have found aesthetically appealing, whether it is seen in a diamond, a flower, or the face of a supermodel. But symmetry isn't always so desirable. Some of the deadliest viruses in the biological books— from influenza to herpes, from polio to the AIDS virus—are constructed using the shape of an icosahedron.

Imagining Shapes

Imagine hanging a cube-shaped decoration on a Christmas tree, with the string attached to one of the corners, or vertices. If you cut the cube horizontally between the point at the top and the point at the bottom, you get two pieces, each with a new face. What is the shape of that new face? The answer is at the end of the chapter.

IS THE BEIJING OLYMPIC SWIMMING CENTER UNSTABLE?

The swimming center built for the Beijing Olympics is an extraordinarily beautiful sight, especially lit up at night, when it looks like a transparent box full of bubbles. Its designer, the firm Arup, was keen to capture the spirit of the aquatic sports played inside but wanted also to give the building a natural, organic look.

The firm began by looking at shapes that can tile a wall—like squares or equilateral triangles or hexagons—but decided that these were all too regular and didn't capture the organic quality the firm was after. It explored other ways in which nature packs things together, like crystals or cell structures in plant tissue. In all these structures, there are examples of the sort of shapes that Archimedes discovered made such good soccer balls, but Arup was particularly drawn to the shapes made by lots of bubbles packed together to make foam.

Considering that it took until 1884 to prove that the sphere is the most efficient shape for a single bubble, it may not come as a surprise that sticking more than one bubble together to make foam leads to some tough questions that are still vexing mathematicians today. If you have two bubbles that contain the same volume of air, what shape do they make when they join together? The rule is always that bubbles are lazy and look for shapes with the least energy. Energy is proportional to surface area, so they try to make a shape that has the smallest surface area of soap film. Since two joined bubbles share a boundary, they can make a shape with smaller surface area than just two bubbles touching.

If you blow bubbles and two bubbles of the same volume fuse together, then the combination looks like this:

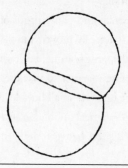

Figure 2.9

The two partial spheres will meet at an angle of 120 degrees and be separated by a flat wall. This is certainly a stable state—if it wasn't, nature wouldn't let the bubbles stay as they are. But the question is whether there might be another shape that has even less surface area—and therefore less energy—that would make it even more efficient. It might require putting some energy into the bubbles to take them out of their current stable state, but perhaps there is an even lower energy state that two bubbles could assume. For example, perhaps the two fused bubbles could be bettered by some weird configuration with less energy in which one bubble takes the shape of a bagel and wraps itself around the other bubble, squeezing it into the shape of a peanut, like this:

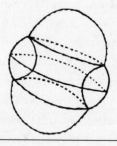

Figure 2.10

The first proof that the fused bubbles couldn't be bettered was announced in 1995. Although mathematicians don't really like asking for help from a computer (because that doesn't appeal to their sense of elegance and beauty), they needed one to check through the extensive numerical calculations that were involved in their proof.

Five years later, a pencil-and-paper proof of the double-bubble conjecture was announced. It actually proved a more general conjecture: if the bubbles do not enclose the same volume, but rather one is smaller than the other, then the bubbles fuse together so that the wall between the bubbles is no longer flat but bent into the small bubble. The wall is part of a third sphere and meets the two spherical bubbles in such a way that the three soap films have angles of 120 degrees between them:

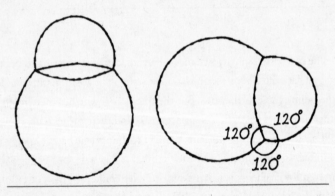

Figure 2.11

In fact, this 120-degree property turns out to be a general rule for the way soap bubbles fuse together. It was first discovered by Belgian scientist Joseph Plateau, who was born in 1801. While he was doing research into the effect of light on the eye, he stared at the sun for half a minute, and by the age of 40, he was blind. Then, with the help of relatives and colleagues, he switched to investigating the shape of bubbles.

Plateau began by dipping wire frames into bubble mixture and examining the different shapes that appeared. For example, when you dip a wire frame in the shape of a cube into the mixture, you get 13 walls that meet at a square in the middle:

It isn't just Arup and the Chinese authorities who are interested in the shape of lots of bubbles squashed together. Understanding the structure of foam helps us to understand much else in nature—for example, the structure of plant cells, the structure of chocolate and whipped cream, and the structure of the head on a pint of beer. Foam is used to put out fires, protect water from radioactive spills, and process minerals. Whether you are interested in fighting fires or making sure that the head on your Guinness doesn't vanish too quickly, the answer lies in understanding the mathematical structure of foam.

WHY DOES A SNOWFLAKE HAVE SIX ARMS?

One of the first people to try to give a mathematical answer to this question was the seventeenth-century astronomer and mathematician, Johannes Kepler. He got his idea for why snowflakes have six arms by looking inside a pomegranate. The seeds in a pomegranate start off as spherical balls. As anyone who sells fruit knows, the most efficient way to fill space with spherical balls is to arrange them into layers of hexagons. The layers fit neatly on top of one another, with each ball nestling over three balls in the layer below. Together, the four balls are arranged so that they are at the corners of a tetrahedron.

Kepler conjectured that this was the most efficient way to pack space— in other words, the arrangement in which the spaces between the balls take up the least volume. But how can you be sure that there isn't some other complicated arrangement of balls that would improve upon this hexagonal packing? The Kepler conjecture, as this innocent statement came to be known, would obsess generations of mathematicians. A proof did not appear until the end of the twentieth century, when mathematicians joined forces with the power of the computer.

Now, back to the pomegranate. As the fruit grows, the seeds begin to squash together, morphing from spheres into shapes that fill space completely. Each seed at the heart of the pomegranate is in contact with 12 others, and as they squash together, the seeds change into shapes with 12 faces. You might think that the dodecahedron with its 12 pentagonal faces would be the shape the seeds adopted, but you can't put dodecahedrons together so that they stack perfectly, filling all the available space. The only Platonic shape that stacks perfectly to fill space is the cube. Instead, the 12 faces on the seed form a kite shape. Called the rhombic dodecahedron, it is a shape often found in nature:

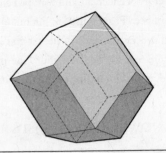

Figure 2.17

Crystals of garnet have 12 faces looking like kites. The word *garnet* actually comes from the Latin for *pomegranate* because the seeds of the fruit also form tiny red 12-sided solids with kite-shaped faces.

Analyzing the kite-shaped faces of the pomegranate seed inspired Kepler to start investigating all the possible symmetrical shapes that could be built out of this slightly less symmetrical kite-shaped face. Plato had considered shapes made from one perfectly symmetrical face; Archimedes took things a step further by looking at shapes made from two or more symmetrical faces. Kepler's investigations sparked off a whole industry dedicated to different shapes that extend the ideas of Plato and Archimedes. We now have the Catalan solids, the Poinsot solids, the Johnson solids, shaky polyhedra, and zonohedra—and many more exotic objects.

Kepler believed that the hexagons at the heart of the way balls pack together were responsible for the snowflake having six arms. His analysis formed the subject of a book he dedicated to an imperial diplomat named Matthaeus Wackher as a New Year's present—an astute move by a scientist always on the lookout for funding. As spherical raindrops froze in clouds, Kepler thought, they were somehow packing themselves together like the pomegranate seeds. A nice idea, but it turned out to be wrong. The real reason for the snowflake's six arms relates to the molecular structure of water—something that would be revealed only with the invention of X-ray crystallography in 1912.

A molecule of water is made up of one oxygen atom and two hydrogen atoms. When water molecules bind together to form crystals, each oxygen atom shares its hydrogen atoms with neighboring oxygen atoms and, in turn, borrows two other hydrogen atoms from other water molecules. So an ice crystal assembles itself with each oxygen atom connected to four hydrogen atoms. In a ball-and-stick model, four balls representing hydrogen atoms are arranged around each oxygen atom in a shape that ensures that each hydrogen atom is as far away as possible from the three other hydrogen atoms. The mathematical solution to such a requirement is to position each hydrogen atom at the corner of a tetrahedron—the Platonic shape made up of four equilateral triangles—with the oxygen atom at its center:

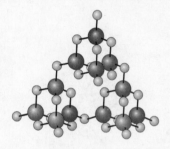

Figure 2.18

The crystal structure that emerges has something in common with a stack of oranges at the grocery store, in which three oranges in one layer have a fourth orange placed on top to make a tetrahedron. But if you look instead at each layer of oranges, you'll see hexagons everywhere. These hexagons, which appear in the ice crystals, are the key to the snowflake's shape. So Kepler's intuition was right—stacking oranges and the six-armed snowflake were related, but it wasn't until we were able to look at the atomic structure of snow that we could see where the hexagons were hidden. As the snowflake grows, the water molecules attach themselves to the six points of the hexagon, building up the six arms of the snowflake.

It is in this passage from the molecular to the large snowflake where the individuality of each snowflake asserts itself. And while symmetry is at the heart of the creation of the water crystal, it is another important mathematical shape that controls the evolution of each flake: the fractal.

HOW LONG IS THE COASTLINE OF BRITAIN?

Is the coastline of Britain 18,000 km long? Or 36,000 km? Or is it even longer? Surprisingly, the answer to this question is far from obvious and is related to a mathematical shape that wasn't discovered until the middle of the twentieth century.

Of course, with the tides coming in and out twice a day, the length of Britain's coastline is constantly varying. But even if we fix the coastline, it's still not clear how long it is. The subtlety arises from the question of how finely you measure the length of the coast. You could start by laying meter rulers end to end and counting how many rulers you need to circumnavigate the country, but using rigid meter rulers is going to miss a lot of smaller-scale detail.

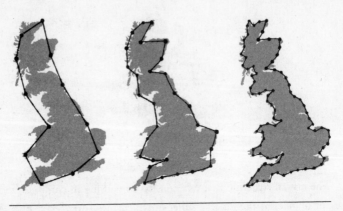

Figure 2.19   Measuring the coastline of Britain.

If you used a long piece of rope instead of rigid rulers, you would be able to follow more of the intricate shape of the coastline. When you pulled the rope out straight to make your measurement, the length of the coastline would be considerably longer than the estimate obtained with meter rulers. But there's a limit to the flexibility of a rope, which can't capture the intricate detail of the contours of the coastline on the centimeter scale. If we used a thin thread, we would be able to capture more of this detail, and then our estimate of the length of the coastline would be greater still.

The Ordnance Survey gives the length of the coastline of Britain as 17,819.88 km. But measure the coastline in finer detail, and you will get double that length. As an illustration of just how difficult it is to pin down geographical lengths, in 1961, Portugal claimed that its border with Spain was 1,220 km, while Spain said it was only 990 km. The same level of discrepancy was found between the borders of Holland and Belgium. In general, it's always the smaller country that calculates the longer border . . .

So, is there any limit to this process? Perhaps the more we zoom in on the detail, the longer the coastline becomes. To show how this is possible, let's build a piece of mathematical coastline. To make the coastline you'll need a ball of string. Start by pulling out 1 m of string from the ball and laying it down on the floor:

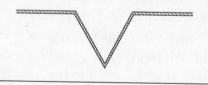

Figure 2.20

This is too straight to be a real coastline, so let's make a large inlet in this straight bit of coast. Pull out some more string from the ball so that the middle third of the string is replaced by two sides of the same length that go in and out:

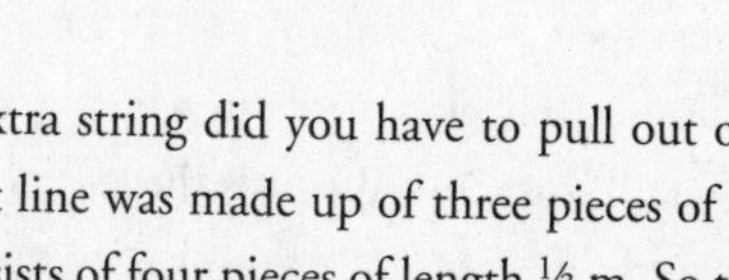

Figure 2.21

How much extra string did you have to pull out of the ball to make this inlet? The first line was made up of three pieces of length ⅓ m, while this new coast consists of four pieces of length ⅓ m. So the new length is ⁴⁄₃ times the first length—that is, ⁴⁄₃ m.

This new coast is still not very intricate. So again, divide each of the smaller lines into three and replace the middle third of each line by two sides of the same length. Now we have this coastline:

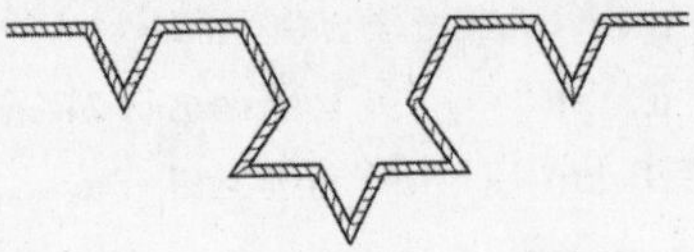

Figure 2.22

How long is this coastline? Well, each of the four lines has again increased in length by a factor of ⅓. So the length of the coast is now ⅓ × ⅓ m = (⅓)² m.

You've probably guessed what we're going to do next. Keep repeating this procedure by dividing the straight lines into three and replacing the middle section with two lines of the same length. Each time we do this, the shape grows in length by a factor of ⅓. If we do this one hundred times, the length of our coastline will have increased by a factor of (⅓)¹⁰⁰, which makes it just over three billion kilometers. Laid out straight, a piece of string that long would stretch from the earth to Saturn.

If we carried out this procedure an infinite number of times, we would get a length of coast that was infinitely long. Of course, physics prevents us from dividing things beyond a certain limit, determined by what is called the Planck constant. This is because, according to physicists, it is actually impossible to measure a distance smaller than 10^{-34} m without creating a black hole that would swallow up the measuring device. When we do our trick of repeatedly adding smaller and smaller inlets to our coastline, by the time we get to the seventy-second step, the length of the lines will already be smaller than 10^{-34} m. But mathematicians are not physicists—we live in a world in which you can divide a line up an infinite number of times and not vanish into a black hole.

Another way to see why the coastline has infinite length is to consider a piece of coastline between two points, A and B, on a coastline that runs from A to E. Let's suppose this piece between points A and B has length L. Assuming that the length from A to B equals the length from B to C, from C to D, and from D to E, if we magnify this piece of the coast from A to B three times, the result is an exact copy of the whole coastline from A to E. So the complete coastline has length $3L$. On the other hand, if we take four copies of this smaller piece, we can put them end to end to cover the complete coastline: A to B, B to C, C to D, and D to E. From this viewpoint, the length of the coast is $4L$, because we need four copies of the small piece to build the coastline. But they have to have the same length, whichever way we measure it. So how can $4L = 3L$? The only resolution to this equation is if L is either of length 0 or has infinite length.

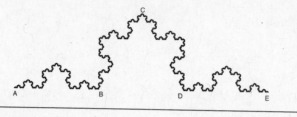

Figure 2.23

This infinite coastline is actually one side of a shape called the Koch snowflake, named after its inventor, the Swedish mathematician Helge von Koch, who constructed it at the beginning of the twentieth century.

This mathematical shape has too much symmetry to look like a real coastline, and doesn't look particularly natural or organic, but if you randomize whether the lines you add each time cut into the coast or jut out into the sea, then things begin to look far more convincing. Here are pictures made by the same procedure as was just described, except that a coin was tossed to decide each time whether the lines were added below or above the line that was removed:

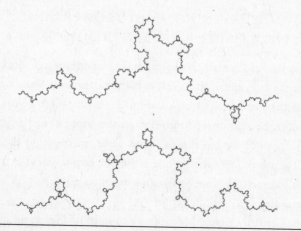

Figure 2.24

If you join several of these coastlines together, you get something that looks remarkably like a medieval map of Britain:

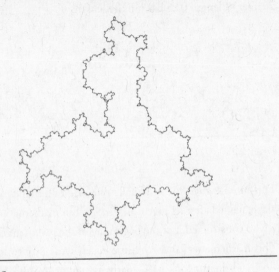

Figure 2.25

So if you are ever asked what the length of the coastline of Britain is, frankly, you can choose whatever answer you like. Isn't that the kind of math question everyone dreams of in school?

WHAT DO LIGHTNING, CAULIFLOWER, AND THE STOCK MARKET HAVE IN COMMON?

In 1960, the French mathematician Benoit Mandelbrot was asked to give a talk to the economics department at Harvard University about his recent work on the distribution of large and small incomes. When he entered his host's office, he was rather perturbed to see the graphs he had prepared for his talk drawn on the blackboard. "How come you've got my data in advance?" he asked. The curious thing is that the graphs weren't anything to do with incomes—they were variations in cotton prices that his host had been analyzing in a previous lecture.

The similarity piqued Mandelbrot's curiosity and led to his discovery that if you took the graphs of various unrelated sets of economic data, they appeared to have a very similar shape. Not only that, but whatever timescale you looked at, the shapes seemed to be the same. For example, variations in cotton prices over eight years looked like variations over eight weeks, and they looked much the same as variations over eight hours.

The same phenomenon occurs with the coastline of Britain. Take, for example, the images in Figure 2.26. They are all sections of the coastline of Scotland. One is from a map with a scale of 1:1,000,000. The others are much more detailed maps—one with a scale of 1:50,000, the other with a scale of 1:25,000. But can you match the pictures to the scale? However much you zoom in or out, these shapes seem to retain the same level of complexity. This isn't true of all shapes. If you draw a squiggly line and zoom in and magnify a portion of it, then at some point it will begin to look quite simple. What characterizes the shape of a coastline or Mandelbrot's graphs is that however far you zoom in, the complexity of the shape is retained.

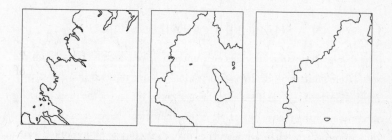

Figure 2.26   The coastline of Scotland at different magnifications. From left to right, original map scales of 1:1,000,000; 1:50,000; and 1:25,000.

As Mandelbrot began to look further afield, he found these strange shapes, which remain infinitely complex at whatever level of magnification you look at them, all over the natural world. If you break off a floret from a cauliflower and magnify it, it looks remarkably like the cauliflower you started with. If you zoom in on a jagged bolt of lightning, then instead of looking quite straight, the magnified section looks like a copy of the original bolt. Mandelbrot christened these shapes fractals, and referred to them as "the geometry of nature" since they represent a genuinely new sort of shape only really recognized for the first time in the twentieth century.

There is a practical reason for the natural evolution of these fractal shapes. The fractal character of the human lung means that even though it sits inside the finite volume of the rib cage, its surface area is huge, and it can therefore absorb a lot of oxygen. The same goes for other organic objects. Ferns, for example, are looking to maximize their exposure to sunlight while not taking up too much space. It harks back to nature's great ability to find shapes with the greatest efficiency. Just as the bubble found that the sphere is the shape that suits its needs best, life-forms have instead gone to the other end of the spectrum, choosing fractal shapes of infinite complexity.

The remarkable thing about fractals is that although they have this infinite complexity, they are actually generated by very simple mathematical rules. At first glance, it is difficult to believe that the complexity of the natural world could be based on simple mathematics, but the theory of fractals has revealed that even the most complex features of the natural world can be created by simple mathematical formulas.

HOW CAN A SHAPE BE 1.26-DIMENSIONAL?

The shapes that mathematicians encountered before fractals appeared on the scene were one-, two-, or three-dimensional—a one-dimensional line, a two-dimensional hexagon, a three-dimensional cube. Yet one of the most amazing discoveries in the theory of fractals is that these new shapes turn out to have dimensions greater than 1 but smaller than 2. If you are feeling brave, here's an explanation of how a shape can have a dimension between 1 and 2.

The trick is to come up with a clever way to capture why a line is one-dimensional while a solid square is two-dimensional. Imagine taking a sheet of transparent graph paper, laying it over a shape, and counting how many squares contain part of the shape. Next, take a sheet of graph paper whose squares are half the size of those on the first piece.

If the shape is a line, the number of squares on the graph paper goes up simply by a factor of 2. If the shape is a solid square, the number of squares goes up by a factor of 4, or 2^2. Each time we halve the size of the grid on the graph paper, the number of squares meeting a one-dimensional shape increases by $2 = 2^1$, while for a two-dimensional shape, the number increases by 2^2. The dimension corresponds to the power of 2.

The curious thing is that if you apply this procedure to the fractal coastline we constructed earlier in the chapter, when we halve the grid size of the graph paper, the number of squares that contain part of the coastline goes up by a factor of approximately $2^{1.26}$. So from this perspective, the dimension of the mathematical coastline we constructed deserves to be called 1.26. We have thus created a new definition of dimension.

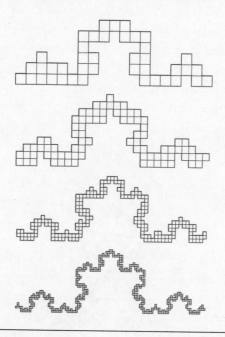

Figure 2.27   How to calculate the dimension of a fractal using graph paper. The dimension measures the increase in the number of pixels as you decrease the size of the pixels.

Instead of graph paper, you can capture these shapes with a pixelated computer screen. Make a pixel black if it contains some of the shape, and leave it white if not. As we increase the screen resolution, the dimension keeps track of the increase in the number of black pixels appearing. For example, if you move from 16 × 16 pixels to 32 × 32, then for a line, the number of black pixels doubles. For a solid square, the number of black pixels increases by a factor of 4, or 2^2. The number of black pixels in a computer image of the Koch snowflake increases by a factor of $2^{1.26}$.

In a sense, the dimension tells us how much this infinite fractal line is trying to fill the space it occupies. If we construct variants of our fractal coastline in which we make the angle between the two lines that we add to the coast smaller and smaller, then the resulting coastline fills more and more of the space. And when we calculate the dimension of each of these sequences of coastline variants, we find that it is creeping closer and closer to 2:

Figure 2.28   As you change the angle of the triangle, the resulting fractal fills more space, and its fractal dimension increases.

If you analyze the fractal dimension of naturally occurring shapes, some interesting things emerge. The coastline of Britain is estimated to have a fractal dimension of 1.25—quite close to·that of the mathematical coastline we constructed. We can think of the fractal dimension as telling us how fast the length of the coastline is increasing as we use smaller and smaller rulers to measure the coast. The fractal dimension of Australia's coastline is estimated to be 1.13, indicating that it is less intricate in some sense than the coastline of Britain. Rather strikingly, the fractal dimension of the coastline of South Africa is only 1.04, which is a sign that it is very smooth. Perhaps the most fractal of all coastlines is Norway's, with all its fjords—it comes out at 1.52.

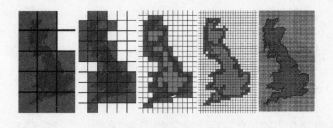

Figure 2.29   What is the dimension of the coastline of Britain?

For objects in three dimensions, we can imagine playing a similar trick, but replacing the graph paper with a mesh of cubes and looking at how the shape intersects these cubes as the mesh gets finer and finer. Cauliflower comes out as a shape with dimension 2.33; a piece of paper crumpled into a ball hits 2.5; broccoli is quite intricate at 2.66; and amazingly, the surface of the human lung has fractal dimension 2.97.

CAN YOU FAKE A JACKSON POLLOCK?

In autumn 2006, a painting by the twentieth-century artist Jackson Pollock became the most expensive ever sold. It was reported that a Mexican financier, David Martinez, paid $140 million for the painting, called simply *No. 5, 1948*.

The painting was created by Pollock's trademark technique of splashing paint across the canvas, which led to his nickname, Jack the Dripper. Critics were shocked at the price that was paid for such a piece, declaring, "Well, I could have made one of those!" and at first sight, it certainly looks as though anyone could splash paint around and hope to become a millionaire. But mathematics has revealed that Pollock was actually doing something more subtle than you might expect.

In 1999, a group of mathematicians led by Richard Taylor of the University of Oregon analyzed Pollock's paintings and discovered that the jerky technique he used actually creates one of the fractal shapes so loved by nature. Magnified sections of a Pollock still look very similar to the full-size version and appear to have the characteristic infinite complexity of a fractal. (Of course, progressively increasing the magnification will eventually reveal the individual spots of paint, but this happens only when you magnify the canvas one thousand times.) The idea of fractal dimension can even be applied to analyze how Pollock's technique developed.

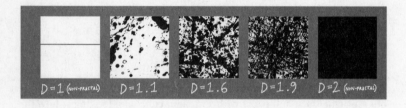

Figure 2.30   The fractal dimension of a painting increases as you splash on more paint.

Pollock started to create fractal pictures in 1943. His early paintings have fractal dimensions in the region of 1.45, similar to the value for the fjords in Norway, but as he developed his technique, so the fractal dimensions crept up, reflecting the fact that the paintings were becoming more complex. One of Pollock's last drip paintings, known as *Blue Poles,* took six months to complete and has a fractal dimension of 1.72.

Psychologists have explored the shapes that people find aesthetically pleasing. We are consistently drawn to images whose fractal dimension is between 1.3 and 1.5, similar to the dimension of many of the shapes found in nature. Indeed, there may be good evolutionary reasons why our brains are drawn to these sorts of fractals, because they are shapes that the brain has become hardwired to recognize as we negotiate the jungle around us. Or maybe, just as the best music sits somewhere between the extremes of boring elevator music and random white noise, these shapes appeal to us because they have a complexity that lies between the too regular and the too random.

If Pollock was creating fractals, how easy is it to replicate his technique? In 2001, a Texas art collector was getting worried that his "Pollock" didn't have a signature or date anywhere on the canvas, so he took it to the mathematicians who'd revealed the fractal dimension of Pollock's style. Their analysis showed that this painting lacked the special fractal character of Pollock's jerky style, so they suggested that it was probably a fake. Five years later, the Pollock-Krasner Authentication Board, set up by the artist's estate to rule on disputed works, asked Richard Taylor and his team to apply their fractal analysis to a collection of 32 paintings that had recently been found in a storage locker and were believed to be by Jackson Pollock. The fractal analysis implied that they, too, were all fakes.

This isn't to say that it's impossible to fake a Pollock—in fact, Taylor has created a piece of equipment he calls the Pollockizer, which paints genuine fractal paintings. Pots containing paint are attached by strings to an electromagnetic coil, which can be programmed to produce chaotic motion, resulting in convincing Pollocks. So although math can help to detect fakes, it can also be used to create images that could well convince the experts.

Fractals are certainly weird shapes if they have dimensions that are not whole numbers—like 1.26 or 1.72—but at least we can draw pictures of them. But things are about to get stranger, because our next step is into hyperspace to explore shapes that exist beyond our three-dimensional world.

HOW TO SEE IN FOUR DIMENSIONS

I can still remember the excitement I felt the day I first "saw" in four dimensions by learning the language that allowed me to conjure up these shapes in my mind's eye. Seeing in four dimensions is possible by using a dictionary, invented by René Descartes, which changes shapes into numbers. He realized that the visual world was often very hard to pin down and wanted a neat mathematical way that would help.

This puzzle shows you that you can't always trust your eyes. As Descartes used to say, "Sense perceptions are sense deceptions":

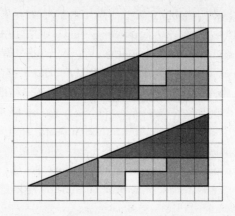

Figure 2.31   Rearrange the pieces and the area appears to decrease by one unit.

Although the second picture is simply the shapes in the first picture rearranged, the total area seems to have been reduced by one block. How can this be? It's because although the hypotenuses of the two small triangles look as if they line up, in fact, they are at slightly different angles—just enough that when you rearrange them, you appear to lose a unit of area.

To deal with this problem of perception, Descartes created a powerful dictionary that translates geometry into numbers, and we are now very familiar with it. When we look up the location of a town in an atlas, we find that it's identified by a two-number grid location. These numbers pinpoint our north–south, east–west location from a point on the equator that lies directly south of Greenwich in London.

For example, Descartes was born in a town in France called . . . Descartes (though when he was born there, it was called La Haye en Touraine), which is at latitude 47° north, longitude 0.7° east. In Descartes's dictionary, his hometown can be described by two coordinates: (0.7, 47).

We can use a similar process to describe mathematical shapes. For example, if I want to describe a square in terms of Descartes's dictionary of coordinates, I can say that it is a shape with four vertices located at the points (0, 0), (1, 0), (0, 1), and (1, 1). Each edge corresponds to choosing two vertices where the coordinates differ in one position. For example, one of the edges corresponds to the coordinates (0, 1) and (1, 1).

The flat two-dimensional world needs just two coordinates to locate each position, but if we also want to include our height above sea level, then we could add a third coordinate. We will need this third coordinate, too, if we want to describe a three-dimensional cube in terms of coordinates. A cube's eight vertices can be described by the coordinates (0, 0, 0), (1, 0, 0), (0, 1, 0), (0, 0, 1), (1, 1, 0), (1, 0, 1), (0, 1, 1), and finally the point farthest from the first corner, located at (1, 1, 1).

Again, an edge consists of two points whose coordinates differ in exactly one position. If you look at a cube, you can easily count how many edges there are. But if you didn't have this picture, you could just count how many pairs of points there are that differ in one coordinate. Keep this in mind as we move to a shape for which we don't have a picture.

Descartes's dictionary has shapes and geometry on one side and numbers and coordinates on the other. The problem is that the visual side runs out if we try to go beyond three-dimensional shapes, since there isn't a fourth physical dimension in which we can see higher-dimensional shapes. The beauty of Descartes's dictionary is that the other side of the dictionary just keeps going. To describe a four-dimensional object, we just add a fourth coordinate that will keep track of how far we are moving in this new direction. So although I can never physically build a four-dimensional cube, by using numbers, I can still describe it precisely. It has 16 vertices, starting at (0, 0, 0, 0), extending to points at (1, 0, 0, 0) and (0, 1, 0, 0), and stretching out to the farthest point at (1, 1, 1, 1). The numbers are a code to describe the shape, and I can use this code to explore the shape without ever having to physically see it.

For example, how many edges does this four-dimensional cube have? An edge corresponds to two points in which one of the coordinates is different. Meeting at each vertex there are four edges, corresponding to changing each of the four coordinates one at a time. So that gives us 16×4 edges—or does it? No, because we've counted each edge twice: once as an edge emerging from the vertex at one of its ends, and again as an edge emerging from the point at its other end. So the total number of edges in the four-dimensional cube is $16 \times \frac{1}{2} = 32$. And it doesn't stop there. You can move into five, six, or even more dimensions and build hypercubes in all these worlds. For example, a hypercube in N dimensions will have 2^N vertices. From each of these vertices, there will be N edges emerging, each of which I am counting twice, so the N-dimensional cube has $N \times 2^{N-1}$ edges.

The math gives you a sixth sense, allowing you to play with these shapes that live beyond the bounds of our three-dimensional universe.

WHERE IN PARIS CAN YOU SEE A FOUR-DIMENSIONAL CUBE?

To celebrate the two hundredth anniversary of the French Revolution, the then president of France, François Mitterrand, commissioned the Danish architect Johann Otto von Spreckelsen to build something special in La Défense, the financial district of Paris. The building would line up with several other significant Paris buildings—the Louvre, the Arc de Triomphe, and Cleopatra's Needle—in what has become known as the Mitterrand perspective.

The architect certainly didn't disappoint. He built a huge arch, called La Grande Arche, which is so large that the towers of Notre Dame could pass through the middle, and weighs a staggering three hundred thousand tonnes. Unfortunately, von Spreckelsen died two years before the arch was completed. It has become an iconic building in Paris, but perhaps less well-known to the Parisians who see it every day is that what von Spreckelsen actually built is a four-dimensional cube in the heart of their capital.

Well, it isn't quite a four-dimensional cube, because we live in a three-dimensional universe. But just as the Renaissance artists were faced with the challenge of painting three-dimensional shapes on a flat two-dimensional canvas, so the architect at La Défense has captured a shadow of the four-dimensional cube in our three-dimensional universe. To create the illusion of seeing a three-dimensional cube while looking at a two-dimensional canvas, an artist might draw a square inside a larger square and then join the corners of the squares to complete the picture of the cube. Of course, it's not really a cube, but it presents the viewer with enough information: we can see all the edges and visualize a cube. Von Spreckelsen used the same idea to build a projection of a four-dimensional cube in three-dimensional Paris, consisting of a small cube sitting inside a larger cube with edges joining the vertices of the smaller and larger cubes. If you visit La Grande Arche and count carefully, you can see the 32 edges that we identified in the previous section using Descartes's coordinates.

Whenever I visit La Grande Arche at La Défense, it is uncanny how there is always a howling wind that seems to suck you through the center of the arch. So serious has this wind become that the designers have had to erect a canopy at the heart of the arch to disrupt the flow of air. It's almost as if constructing a shadow of a hypercube in Paris has opened up a portal to another dimension.

There are other ways to get a feel for the four-dimensional cube in our three-dimensional world. Think of how you would make a three-dimensional cube from a piece of two-dimensional card stock. First, you would draw six squares connected in a cross-shape—one square for each face of the cube. Then, you would wrap the cross-shape up to form a cube. The two-dimensional card stock is called the "net" of the three-dimensional shape. In a similar fashion, it is possible in our three-dimensional world to build a three-dimensional net, which, if you had a fourth dimension, could be wrapped up to make a four-dimensional cube.

You could set about making a four-dimensional cube by cutting out and assembling eight cubes. These will be the "faces" of your four-dimensional cube. To make the net of the four-dimensional cube, you need to join these eight cubes together. Start by gluing together the first four cubes into a column, one stacked upon the other. Next, take the remaining four cubes and stick them to the faces of one of the four cubes in the column. Your unwrapped hypercube should now look like two intersecting crosses, as shown in Figure 2.32.

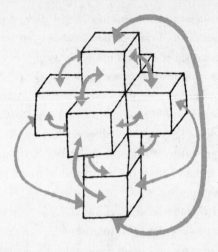

Figure 2.32   How to make a four-dimensional cube from eight three-dimensional cubes.

To fold this thing up, you would need to start by joining the bottom and top cubes in the column. The next step would be to join the outward-facing squares of two of the cubes stuck on opposite sides of the column to the bottom cube in the column. Then finally, you'd need to glue the faces of the other two side-cubes to the remaining two faces of the bottom cube. The problem is, of course, that as soon as you start to fold this thing together, you get into a tangle, as there just isn't enough room in our three-dimensional world. You need a fourth dimension in which to wrap it up as I have described.

Just as the architect in Paris was inspired by the shadow of the four-dimensional cube, so the artist Salvador Dalí was intrigued by the idea of this unwrapped hypercube. In his painting *Crucifixion (Corpus Hypercubus)*, Dalí depicts Christ crucified on the three-dimensional net of a four-dimensional cube. For Dalí, the idea of the fourth dimension as something beyond our material world resonated with the spiritual world beyond our physical universe. His unwrapped hypercube consists of two intersecting crosses, and the picture suggests that Christ's ascension to heaven is connected with trying to wrap this three-dimensional structure into a fourth dimension, transcending physical reality.

However we try to depict these four-dimensional shapes in our three-dimensional universe, they can never give a complete picture, just as a shadow or silhouette in the two-dimensional world can give only partial information. As we move and turn the object, the shadow changes, but we never see everything. This theme was picked up by novelist Alex Garland in his book *The Tesseract*, which is another name for a four-dimensional cube. The narrative describes different characters' views of the central story set in the gangster underworld of Manila. No single narrative provides a complete picture, but by piecing together all the strands, like looking at the many different shadows cast by a shape, you start to understand what the story might be. But the fourth dimension is not just important for constructing buildings, paintings, and narratives. It might also be the key to the shape of the universe itself.

IN THE COMPUTER GAME ASTEROIDS, WHAT SHAPE IS THE UNIVERSE?

In 1979, the computer arcade company Atari released its most popular game, Asteroids. The object of the game was to shoot and destroy asteroids and flying saucers while trying to avoid colliding with passing asteroids or being shot by the saucers' counterfire. The arcade version was so successful in the United States that many arcades had to fit bigger cash boxes to hold all the quarters that were being fed into the machines.

But it is the geometry of the game that is interesting from a mathematical point of view: as soon as the spaceship flies off the top of the screen, it magically reappears at the bottom. Similarly, if you exit the screen on the left, the spacecraft reappears, entering on the right of the screen. What is happening is that our spaceman is stuck in a two-dimensional world in which the entire universe can be seen on the screen. Although this is a finite universe, it has no boundaries. Because the spaceman never hits an edge, he isn't living in a rectangle, but is flying around in a more interesting universe. Can we work out what shape his universe is?

If the spaceman exits the screen at the top and reenters at the bottom, then these bits of his universe must be connected. Imagine that the computer screen is made out of flexible rubber, so that we can bend it around and join the top to the bottom. As the spaceman flies vertically, we can now see that he is actually just travelling around and around a cylinder.

What about the other direction? When he exits the screen at the left, he enters again at the right, so the two ends of the cylinder must also be connected. If we mark the points where they are connected, we find that we must bend the cylinder around and join its top to its bottom. So actually, our spaceman is living on a bagel, or what we mathematicians call a torus.

What I've illustrated with this piece of rubber is actually a new way in which mathematicians started to look at shapes about a hundred years ago. For the ancient Greeks, geometry (a word that comes from the Greek and means literally "measuring the earth") was about calculating distances between points and angles. But analyzing the shape of the spaceman's universe in the game of Asteroids is not so much about the actual distances in our spaceman's universe but about how it is all connected. This new way of looking at shapes, in which I'm allowed to push and pull them around as if they were made from rubber or Plasticine, is called topology.

Many people use topological maps every day. Although geometric maps are geographically accurate, they're not very good for finding your way around. Londoners, for instance, use a topological map for finding their way around the Underground. This topological map was first designed in 1933 by Harry Beck; he pushed and pulled the geometric map of London to get something that was much more user-friendly and is now familiar around the world.

Understanding whether a knot can be untangled is also a question of topology, because we are allowed to pull the ropes around but not cut them. This is of fundamental importance to biologists and chemists because human DNA tends to fold up into strange knots. Some diseases, such as Alzheimer's, might be related to the way DNA knots itself, and math has the potential to unlock these mysteries.

At the beginning of the twentieth century, the French mathematician Henri Poincaré began to wonder how many topologically different surfaces there are. This is like looking at all the possible shapes that our two-dimensional Atari spaceman might be able to inhabit. Poincaré was interested in these universes from a topological perspective, so two universes should be regarded as the same if one universe can be morphed into another continuously and without making any cuts. For example, the two-dimensional surface of a sphere is topologically the same as the two-dimensional surface of a rugby ball because one can be molded into the other. But this spherical universe is a different topological shape to the torus in which Atari's spaceman is flying around, because you can't morph a sphere into a doughnut without cutting or gluing the shape. But what other shapes are out there?

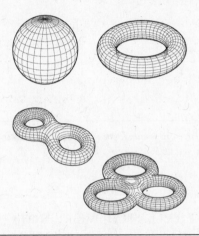

Figure 2.33   The first four shapes in Henri Poincaré's topological classification of how to wrap up two-dimensional surfaces.

Poincaré was able to prove that, however complicated a shape might be, it is always possible to continuously morph it into one of the following shapes: a sphere or a torus with one hole, two holes, three holes, or any finite number of holes. From a topological point of view, this is a complete list of possible universes for our Atari spaceman. It is the number of holes—what mathematicians refer to as the genus—that characterizes the shape. For example, a teacup is topologically identical to a bagel because both have one hole. A teapot has two holes—one in the spout and one in the handle—and can be molded to look like a pretzel with two holes in it. It is perhaps more challenging to understand why the shape in the following figure, which also has two holes, can be morphed into the two-holed pretzel. With the bagels interlocked, it looks like you'd have to cut the shape to morph it successfully, but you don't. At the end of the chapter, I explain how to undo the rings without any cutting.

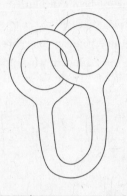

Figure 2.34 *How can you undo the two interlocked rings by continuously morphing them and without cutting them?*

HOW CAN WE TELL THAT WE'RE NOT LIVING ON A BAGEL-SHAPED PLANET?

In ancient times, it was assumed that the earth was flat. But as soon as people started to travel great distances, the question of the large-scale shape of the earth became more important. If the world were flat, then, it was agreed, if you travelled far enough you would fall off the edge—unless you never reached the edge because the world went on forever.

Many cultures began to realize that the earth was most probably curved and finite. The most obvious proposal for the shape is, of course, a ball, and several ancient mathematicians gave incredibly accurate calculations for the size of this ball based only on analyzing how shadows changed throughout the day. But how could scientists be so sure that the surface of the earth wasn't wrapped up in some other interesting shape? How could they tell that we weren't living on, for example, the surface of some huge bagel, rather like the Asteroids spaceman stuck in his two-dimensional, bagel-shaped universe?

One way is to go on an imaginary journey in these alternative worlds. So let's set down an explorer on the surface of a planet and tell him that it is either a perfect sphere or a perfect bagel shape. How can he distinguish between the two? We get him to head off in a straight line across the surface of the planet with a brush and a bucket of white paint, which he uses to mark out his path. Eventually, the explorer will return to where he started, having traced his path as a huge white circle round the planet.

We now give him a bucket of black paint and tell him to head off in another direction. On a spherical earth, for whatever direction he chooses, the black path will always cross the white path before he gets back to his starting point. Remember that the explorer is always travelling in a straight line on the surface. The point where the two paths cross will be at the "pole" opposite the point where the explorer started.

Figure 2.35   *Two paths on a sphere cross at two places.*

On the surface of a bagel-shaped planet, things are rather different. The white journey could take him around the inside of the bagel, through the hole, and out the other side. But if on his black journey he set off at 90 degrees to the white path, he'd walk around the hole without going inside it. So it is possible to make two journeys that will meet only at the place you started from.

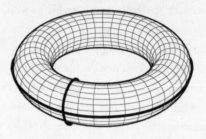

Figure 2.36   There are paths on a torus that cross only once.

The problem is that the surface of a planet is not generally perfectly spherical or bagel-shaped—it is distorted. Dents in the surface where meteorites have struck will have bent it out of shape, so as the explorer travels in a straight line, hitting one of these dents or bumps would send him off in a new direction. In fact, it's quite possible that if the explorer heads off in a straight line, he will never return to his starting point. Since the dented shapes are still just distorted versions of the sphere or bagel, perhaps there are other ways to distinguish them. This is where the subject of topology is so powerful, because it's concerned not so much with the shortest path between two points, but with whether one path can be molded into another.

Let's now send our explorer off with a white elastic rope, which he lays down behind him. He keeps going until he comes back to the beginning again and then joins the ends of the rope so that he has a noose around the planet. He then heads off in a new direction with a black elastic rope and again keeps going until he returns to his starting point. If the planet is essentially a ball or sphere with a few dips and peaks in it, then without cutting either rope, he can always move the black rope so that it lies completely over the white rope. But on a bagel-shaped planet, this isn't always possible. If the black rope is wrapped through the inside of the hole of the bagel and the white rope is laid down on a circle going around the outside ring of the bagel, then there is no way to pull the black rope to match up with the white rope without cutting it. So the explorer can tell whether there's a hole in the planet by making journeys around it and without ever leaving the surface to find out what shape it is.

Here are two other curious ways to tell whether you are on a ball-shaped planet or a bagel-shaped planet. Imagine that both planets are covered in fur. The explorer on the furry bagel will find that he can comb the hair so that it all lies down smoothly—for example, by combing the hair into the hole and back out the other side. But the explorer on the furry ball is going to be in trouble. However hard he tries to comb the hair on the ball-shaped planet, there will always be a crown where the hair sticks up.

Bizarrely, this has a strange implication for the weather on these two planets, as the hairs can be thought of as the direction in which the wind is blowing on these different worlds. On the globe, there is always somewhere where there is no wind blowing—at the crown—but on the bagel, it is possible for the wind to be blowing everywhere on the surface.

Another difference between these two planets is in the maps that can be drawn on them. Divide each planet into different countries, and then try to color the maps so that no two countries with a common border have the same color. On the surface of the spherical earth, you can always get away with just four colors. On a map of Europe, look at the way Luxembourg is boxed in by Germany, France, and Belgium—you can see that you need at least four colors. But the extraordinary thing is that you don't need any more colors—there is no way to redraw the boundaries of Europe that will force you to invest in a fifth color. However, to prove this was no easy matter. The proof that there really wasn't some crazy map that would need a fifth color was one of the first in mathematics that had to resort to employing a computer to check several thousand maps—coloring all those maps by hand would have taken too long.

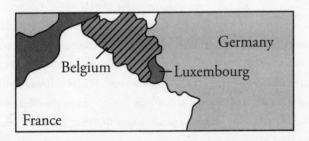

Figure 2.37   Four colors are needed to color a map of Europe.

What about the cartographers living on the bagel-shaped planet—how many buckets of paint are they going to need? There are maps on the bagel-shaped planet that actually need as many as seven colors. Remember from the Asteroids game that we can wrap up the rectangular screen to make a bagel in which the top and bottom are joined to make a cylinder, and then the left- and right-hand sides of the screen that make up the ends of the cylinder are joined to make the bagel. Here's a map on the surface of the bagel before it is joined up; it needs seven colors once it has been sewn together.

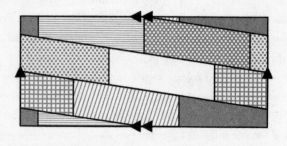

Figure 2.38   Wrap up this map into the shape of a bagel by joining the top and the bottom and then joining the two ends. You'll find that you need seven colors to color it.

And now, having journeyed through the mathematics of bubbles and bagels, and fractals and foam, we are ready to tackle the ultimate question of the mathematics of shape.

WHAT SHAPE IS OUR UNIVERSE?

This is one question that has obsessed humankind for millennia. The ancient Greeks believed that the universe was bound by a celestial sphere on the inside of which the stars were painted. This sphere would rotate every 24 hours, which explained the movement of the stars. However, there is something rather unsatisfying about this model: if we travelled out into space, would we eventually hit a wall? And if so, what is on the other side of that wall?

Isaac Newton was one of the first to propose that perhaps the universe has no boundary—that it is infinite. But there is a third possibility that the universe is finite yet has no boundary. How is this possible?

This is a similar problem to the one faced by our explorers on a world that has a finite surface area but no edges or boundaries. But instead of being stuck on a two-dimensional surface, we are inside a three-dimensional universe. Is there an elegant way to find the shape of this universe and resolve the apparent paradox of it having no boundary yet still being finite?

It took the invention of four-dimensional geometry in the middle of the nineteenth century for shapes to appear that provided a possible answer. Mathematicians realized that the fourth dimension gave them the room to wrap up our three-dimensional universe to create shapes that are finite in volume yet have no boundaries, just as the two-dimensional surface of the earth or the surface of a bagel is finite in area but has no edges.

We have already seen how a finite two-dimensional universe like the Asteroids universe is actually the surface of a three-dimensional bagel, but we are three-dimensional travellers who can travel in a third dimension. Perhaps the universe we live in behaves the same way as the universe in Asteroids. To start with, imagine freeze-framing the universe just after the big bang, when it has expanded to the size of your bedroom. This bedroom-sized universe is finite in volume, but it doesn't have any boundaries—because the bedroom is connected together in a rather curious way.

Imagine that you're standing in the middle of the bedroom facing a wall. (I'm assuming that your bedroom is cube-shaped.) As you walk forward, instead of hitting the wall in front of you, you actually emerge through the wall that was behind you. Similarly, passing through the wall behind you sees you emerge through the wall in front. If you change direction by 90 degrees and head toward the wall on your left, then as you pass through it you emerge through the wall on the right, and vice versa. So the way we have connected up your bedroom is just like in the game of Asteroids.

But we are three-dimensional space travelers, and there's a third direction we can head in. When we fly up into the ceiling, instead of bouncing off of it we pass through it and find ourselves emerging through the floor. Travelling in the opposite direction will take us out through the floor and back in through the ceiling.

The shape of this universe is actually the surface of a four-dimensional bagel, or hyperbagel, but just like the spaceman who is trapped in the game of Asteroids and can't get out of his two-dimensional world to see how his universe is wrapped up, we can never see this hyperbagel. But by using the language of mathematics, we can still experience its shape and explore its geometry.

Our universe has now expanded way beyond the size of a bedroom, but it might still be stuck together like a hyperbagel. Think about the light that travels out in a straight line from the sun. Maybe, rather than disappearing into infinity, this light can loop around and come back and hit the earth. If so, then one of the distant stars out there could be our sun seen from the opposite side, since the light would have travelled all the way around the hyperbagel and finally back to earth. We could therefore be staring at our own sun when it was much younger.

This seems incredible, but just think of sitting in your mini bagel–bedroom universe and striking a match. Looking toward the wall in front, you see the light of the match right in front of you. Now turn around. Looking at the back wall of your bedroom, you will see the match again, only a little farther off because the light from the match heads toward the wall that was in front of you and then reemerges through the back of the bedroom and hits your eye.

Instead of a hyperbagel, we could be living on the surface of a four-dimensional soccer ball. Some astronomers believe that we might be living on a shape that looks like a 12-faced dodecahedron where, as in your mini bedroom universe, when you hit one of the faces of the dodecahedron, you reenter the universe on the opposite face. Perhaps we have come full circle and returned to the model Plato proposed two thousand years ago, according to which our universe was enclosed in some kind of glass dodecahedron with stars stuck on its surface. Maybe modern mathematics can make sense of this model, where the faces of this shape are joined up to make a universe with no glass walls.

But are there other shapes that the universe could be? Remember that Poincaré classified all the possible shapes that a two-dimensional surface, like the surface of our planet, could have. The surface can be wrapped up like a soccer ball; a bagel; or a pretzel with two holes, three holes, or more holes. Poincaré proved that any other shape that you might try to make can be morphed into a ball or a pretzel with holes in.

So what about our three-dimensional universe—what shape could it be? Called the Poincaré conjecture, this is the million-dollar problem for this chapter. It is rather special because in 2002, news emerged that the Russian mathematician Grigori Perelman had solved it. His proof has been checked by many mathematicians, and it is now acknowledged that he has indeed classified all the possible shapes the universe might be. This was the first of the million-dollar problems to be solved, but when Perelman was offered the million dollars in June 2010, he amazingly decided to turn it down. For him, the prize wasn't money but solving one of the biggest problems in the history of mathematics. He had already turned down a Fields medal, the mathematicians' equivalent of a Nobel prize. In this age of celebrity and materialism, there is something rather noble about a man who gets his kicks out of solving theorems and not winning prizes.

With the acceptance of Perelman's proof, mathematicians have sorted out the possible shapes there could be. Now it is up to astronomers to look into the night sky and pin down which one best describes the elusive shape of the universe.

SOLUTIONS

Imagining Shapes

The slice cuts all six faces, and each face contributes an edge to the new face. The shape has to be symmetrical, so you get a hexagon.

Unlinking the Rings

This is how to undo the two interlocked rings by continuously morphing them into a double-holed torus:

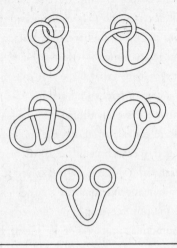

Figure 2.39

Three

THE SECRET OF THE WINNING STREAK

Playing games is an essential part of the human experience. Games are a safe way of exploring real-life situations. Monopoly is a microcosm of the economy, chess is an 8 × 8 battlefield, and poker is an exercise in assessing risk. Games allow us to develop ways of predicting how, given certain rules, events will unfold and to plan accordingly. They teach us about chance and unpredictability, which play such essential parts in nature's game of life.

From ancient civilizations all around the world, we have a fascinating assortment of games: stones thrown in the sand, sticks tossed in the air, tokens placed in hollows carved into wooden blocks, hands used to compete, and pictures drawn on cards. From ancient mancala to Monopoly, from the Japanese game of go to the poker tables of Vegas, games are invariably won by whoever is best at taking a mathematical, analytical approach. In this chapter, I will show you how math is the secret to the winning streak.

HOW TO BECOME THE ROCK-PAPER-SCISSORS WORLD CHAMPION

Jan-ken-pon in Japan, *ro-sham-bo* in California, *kai-bai-bo* in Korea, and *ching-chong-cha* in South Africa—the game of rock-paper-scissors is played all around the world.

The rules are very simple: on a count of three, each player makes his or her hand into one of three shapes: a fist for a rock, a flat hand for paper, or two fingers in a V for scissors. Rock beats scissors, scissors beats paper, and paper beats rock. Two of the same is a draw.

Now, the rationale for the first two wins is clear: rock blunts the scissors, scissors cut the paper. But why does paper beat the rock? A sheet of paper isn't much protection against someone throwing a rock at you. But it may be that this convention goes back to ancient China, in the days when a petition to the emperor was symbolized by a rock. The emperor would indicate whether he'd accepted the petition by placing a piece of paper above or below the rock. If the rock was covered by the paper, the petition was refused and the petitioner defeated.

The origins of this game are hard to trace. There is evidence that it was played in the Far East and by Celtic tribes, and even possibly as far back as the ancient Egyptians, who used to play finger games. However, all these cultures seem to have been beaten to the discovery by a group of lizards that have been playing the game in the fight for survival long before humankind was making fists.

The west coast of the United States is home to a species of lizard called *Uta stansburiana,* more commonly known as the common side-blotched lizard. The male comes in three different colors—orange, blue, and yellow—and each color has a different mating strategy. Orange lizards are the strongest and will attack and beat blue lizards. The blue lizards are bigger than the yellow lizards and are happy to engage in battle with them and beat them. But though the yellow lizards are smaller than the blue and orange males, they look like female lizards, and that confuses the orange lizards. So the orange lizards, who are looking for a fight, don't notice the yellow lizards slipping under their gaze and mating with the females. The yellow lizards are sometimes referred to as "sneakers" for the devious way in which they outflank the orange ones. So orange beats blue, blue beats yellow, and yellow beats orange—an evolutionary version of rock-paper-scissors.

Figure 3.1

These lizards have been playing the game in the course of perpetuating their genes for a long time, and it would be interesting to know whether they have developed a strategy for winning. Their population tends to follow a six-year cycle in which first orange dominates, then yellow, then blue, then orange once again. The pattern that emerges is precisely the one that people will use in trying to win the game in one-to-one combat. See too many rocks being thrown and you start to offer paper, but once your opponent sees the run of paper beating the stone, he or she wises up and switches to scissors to cut off your paper run. You soon pick up your opponent's change of behavior and shift to rock.

At its heart, winning this game is all about spotting patterns, and that's a very mathematical trait. If you can predict what your opponent is going to do next because of a pattern of behavior he or she has established, then you're in. The problem is that you don't want there to be any immediately obvious rhythm in the way you respond, or your opponent will gain the upper hand. A huge amount of psychology is going on as each contestant tries to spot patterns in his or her opponent's play, each second-guessing what the other might do next.

Rock-paper-scissors has recently grown from a playground game to an international contest. Each year, a prize of $10,000 awaits the winner of the coveted title of rock-paper-scissors world champion. The roll of honor has been dominated by contestants from North America, but in 2006, Bob "The Rock" Cooper from North London held his nerve to take the title. His training for the tournament? "Several hours of hard practising in front of the mirror each day." I guess this helps to build up the psychology of dealing with your opponent reading your mind. And the secret to his success? His nickname makes other players assume that he'll throw a rock more often than not, which allows him to cut in with scissors to beat the paper that they try to catch him with. But once they've seen through this ruse, Bob "The Rock" uses a more mathematical approach.

From a mathematical, rather than a psychological, point of view, your best strategy is to make your choices random. Your opponent then has nothing at all to go on, because in a truly random sequence, what has gone before will not influence what follows. If I toss a coin ten times, the first nine tosses will have no influence on the outcome of the last toss. Even if you've tossed nine heads, that doesn't mean that the tenth has got to be a tail to balance things out. A coin has no memory.

The strategy of randomizing things gives you only an evens chance of winning, as it makes playing a game of rock-paper-scissors no different from tossing a coin to see who wins. But if I were going up against the world champion, I'd take any strategy that gave me an evens chance of winning. I can't think of many sports in which you can devise a strategy to give you a fifty-fifty chance of beating the world champion. The one hundred–meter sprint? I don't think so.

But how do you come up with a sequence of choices that you can be certain is random and doesn't have some hidden pattern? It's a real problem: we humans are notoriously bad at producing random sequences—we are so addicted to patterns that we tend to let structure seep into any random sequence we try to put together. To help you win the game, you can download a PDF from the Number Mysteries website containing a rock-paper-scissors die you can assemble and use to help you randomize your choices.

Scissors and Cezanne

The game of rock-paper-scissors has been used to decide disputes from fights in the playground to battles in the boardroom. Two auction houses, Sotheby's and Christie's, famously settled a dispute over the right to auction a collection of impressionist paintings by Cézanne and Van Gogh by a single game of rock-paper-scissors.

Each auction house was given the weekend to come up with its choice of play. At great expense, Sotheby's hired a team of top-ranking analysts to produce a winning strategy. The analysts concluded that because it's a game of chance, a random choice is as good as any. So they went with paper. Christie's simply asked the 11-year-old daughter of an employee what she would do. "Everyone always assumes you'll do rock, so they'll choose paper. So go with scissors," she told them. Christie's won the contract.

It just goes to show that math won't always give you the edge.

HOW GOOD ARE YOU AT RANDOMNESS?

Our intuition is generally very bad at realizing the consequences of randomness. I'm going to offer you a bet. I'm going to toss a coin ten times. You give me $1 if there's a run of three heads or three tails. If there isn't, I'll give you $2. Would you take the bet?

What if I upped what I offer to pay you to $4? I reckon that if you were unsure the first time, you would probably now take me on. After all, how likely is it that you'd get three heads or three tails in a row in ten tosses of a coin? Amazingly, you will get such a run over 82 percent of the time. So even if I'm paying out at $4 for no sequence of three, in the long run, I'll still be making a profit.

The exact probability of tossing a coin ten times and getting a run of three heads or three tails in a row is $846/1{,}024$. Here are the gory details of how to calculate this probability. Strangely enough, the Fibonacci numbers we met in chapter 1 are the key to working out the chances—yet more confirmation that these numbers are everywhere. If I toss a coin N times, there are 2^N ways the coin could land. We'll say that g^N is the number of combinations with no runs of three heads or three tails: these are the combinations in which you will win the bet. We can calculate g^N by using the rule for the Fibonacci numbers:

$$g^N = g^{N-1} + g^{N-2}$$

To get the numbers going, you just need to know that $g^1 = 2$ and $g^2 = 4$, because after one or two tosses, no combination can have three heads or three tails as we haven't tossed the coin three times yet. So the sequence goes 2, 4, 6, 10, 16, 26, 42, 68, 110, 178, . . . There are therefore $1,024 - 178 = 846$ different ways that ten coins can land and have a run of three heads or three tails. So the probability is $846/1,024$, roughly 82 percent, that such a run would occur—and that I would win.

Why is the Fibonacci rule the key to calculating g^N? Take all the combinations of $N-1$ tosses with no runs of three heads or three tails. There are g^{N-1} of these. Now make the Nth toss the opposite of the $(N-1)$th toss. Next, take all the combinations of $N-2$ tosses with no runs of three heads or three tails. There are g^{N-2} of these. Now make the $(N-1)$th and Nth tosses the opposite of the $(N-2)$th toss. In this way, you generate all the combinations of N tosses with no runs of three heads or three tails.

HOW CAN I WIN THE LOTTERY?

This is the question I get asked the most when I say that I spend my life playing with numbers. But just like tossing a coin, the numbers that have come up in the previous weeks' draws cannot influence the numbers that come up next Saturday. That's what it means to be random, but some people will never be convinced.

Italy's state-run biweekly lottery draw takes place in ten cities across the country, and participants have to choose numbers from 1 to 90. At one point, the number 53 ball had refused to appear in Venice after nearly two years of lottery draws. Surely, after so long, it was certain to come up the following week—or so thought many Italians. One woman bet all her family's savings on 53 coming up. When it failed yet again to appear, she drowned herself in the sea. Even more tragically, a man shot his family and then himself after running up huge debts betting on the certainty of 53 appearing. It is estimated that Italians invested $3.8 billion—an average of $240 per family—on 53 being a winner.

There were even calls for the government to ban 53 from the draw to put an end to the country's obsession with the number. When the dam finally burst on February 9, 2005, and the 53 ball popped up in the draw, $640 million was paid out to an unspecified number of lottery players. Inevitably, some people accused the state of deliberately withholding the 53 ball to avoid a huge payout, and it wasn't the first time such a rumor had circulated. In 1941, the number 8 ball had failed to appear after 201 draws in Rome. Many believed that Mussolini had fixed its nonappearance and was siphoning off the nation's bets on the 8 ball to help finance Italy's war effort.

Now, to see how lucky you are, let's play our own little lottery. I can't promise you millions of dollars, but the good news is that this lottery is free to enter. To play Number Mysteries lotto, start by choosing six numbers from the 49 numbers on this ticket:

[1]	[2]	[3]	[4]	[5]	[6]	[7]	[8]	[9]	[10]
[11]	[12]	[13]	[14]	[15]	[16]	[17]	[18]	[19]	[20]
[21]	[22]	[23]	[24]	[25]	[26]	[27]	[28]	[29]	[30]
[31]	[32]	[33]	[34]	[35]	[36]	[37]	[38]	[39]	[40]
[41]	[42]	[43]	[44]	[45]	[46]	[47]	[48]	[49]	

Figure 3.2

Chosen your numbers? To see whether you've won, go to the website www.random .org/quick-pick/, or use your smartphone to scan this code.

Select 1 ticket, United Kingdom, National Lottery, and click on "Pick Tickets." If you don't have Internet access, there is a predetermined choice of six numbers at the end of this chapter. Don't cheat, though. Like solving a math puzzle, it's much more fun to get the answer right yourself rather than looking it up.

What are your chances of picking all six numbers correctly and winning the lottery? To calculate the odds, you need to work out how many different possible choices of six numbers there are—call it N. Then the odds that you've chosen the winning numbers are 1 in N. As a warm-up, let's start by looking at how many different ways there are to pick two numbers. There are 49 choices for your first number. For the second, you now have a choice of 48 numbers. Each choice of the first number can be paired with one of the remaining 48 numbers. So that gives us 49×48 possible pairs of numbers. But hold on—we've actually counted each choice twice. For example, if you chose 27 as your first number and then 23 as your second, that's the same as if you'd chosen 23 first and then 27. So there are only half as many pairs of numbers as we first thought, which means that the number of pairs of numbers you can choose is $\frac{1}{2} \times 49 \times 48$.

Now to six numbers. There are 49 choices for the first number, 48 for the second, 47 for the third, 46 for the fourth, 45 for the fifth, and finally 44 choices for the last number. So that's $49 \times 48 \times 47 \times 46 \times 45 \times 44$ combinations of six numbers—except that again, we've counted some combinations more than once. How many times, for example, have we counted the combination 1, 2, 3, 4, 5, 6? Well, we could have chosen any of these six as our first number (say, 5). That would leave five numbers as possible second choices (say, 1), four numbers for the next choice (say, 2), three for the next (say, 6), and two choices for the penultimate number (say, 4). The final number, then, has to be the one that's left (in this case, 3). So we could have picked the six numbers 1, 2, 3, 4, 5, 6 in $6 \times 5 \times 4 \times 3 \times 2 \times 1$ different ways. This is true for any combination of six numbers. So we need to divide $49 \times 48 \times 47 \times 46 \times 45 \times 44$ by $6 \times 5 \times 4 \times 3 \times 2 \times 1$

to get the total possible number of ways to fill out the lottery ticket. The answer? 13,983,816.

This number also tells us your chance of winning, since it's the total number of possible combinations of the ways the balls come out of the lottery machine. In other words, the chance of you choosing the correct combination in the total number of possible combinations is 1 in 13,983,816.

What are the chances that you got no numbers correct? We work it out the same way. Your first number has to be one of the 43 numbers that don't get drawn, your second number one of the remaining 42, and so on. This gives $43 \times 42 \times 41 \times 40 \times 39 \times 38$ different combinations. But each combination has been counted $6 \times 5 \times 4 \times 3 \times 2 \times 1$ times. So the total number of combinations that have no numbers correct is $43 \times 42 \times 41 \times 40 \times 39 \times 38$ divided by $6 \times 5 \times 4 \times 3 \times 2 \times 1$, or 6,096,454. So just under half the number of possible choices will have no numbers that match the winning numbers. To calculate your chance of not getting any numbers right, divide 6,096,454 by 13,983,816. That gives approximately 0.436, or a 43.6 percent chance that you've drawn a complete blank.

So you have a 56.4 percent chance of getting at least one number right. What are the chances of getting two numbers right? To calculate this, you need to find the number of combinations with two correct numbers. You have six choices for the first correct number and another five for the second number. That's 6×5, but again you need to divide by 2 to correct for counting things twice. For the four numbers you get wrong, you have a choice of $43 \times 42 \times 41 \times 40$, which you need to divide by $4 \times 3 \times 2 \times 1$, which is the number of ways you've counted things twice. So the number of combinations with exactly two numbers right is

$$\left(\frac{6 \times 5}{2}\right) \times \left(\frac{43 \times 42 \times 41 \times 40}{4 \times 3 \times 2 \times 1}\right) = 1,851,150$$

Number of correct numbers N	Number of combinations with N correct numbers	Probability of getting exactly N correct numbers
0	$\dfrac{43 \times 42 \times 41 \times 40 \times 39 \times 38}{6 \times 5 \times 4 \times 3 \times 2 \times 1} = 6{,}096{,}454$	$\dfrac{6{,}096{,}454}{13{,}983{,}816} = 0.436$ Nearly 1 in 2
1	$6 \times \dfrac{43 \times 42 \times 41 \times 40 \times 39}{5 \times 4 \times 3 \times 2 \times 1} = 5{,}775{,}588$	$\dfrac{5{,}775{,}588}{13{,}983{,}816} = 0.413$ About 2 in 5
2	$\dfrac{6 \times 5}{2} \times \dfrac{43 \times 42 \times 41 \times 40}{4 \times 3 \times 2 \times 1} = 1{,}851{,}150$	$\dfrac{1{,}851{,}150}{13{,}983{,}816} = 0.132$ About 1 in 8
3	$\dfrac{6 \times 5 \times 4}{2 \times 3} \times \dfrac{43 \times 42 \times 41}{3 \times 2 \times 1} = 246{,}820$	$\dfrac{246{,}820}{13{,}983{,}816} = 0.0177$ About 1 in 57
4	$\dfrac{6 \times 5 \times 4 \times 3}{2 \times 3 \times 4} \times \dfrac{43 \times 42}{2 \times 1} = 13{,}545$	$\dfrac{13{,}545}{13{,}983{,}816} = 0.000969$ About 1 in 1,032
5	$\dfrac{6 \times 5 \times 4 \times 3 \times 2}{2 \times 3 \times 4 \times 5} \times 43 = 258$	$\dfrac{258}{13{,}983{,}816} = 0.0000184$ About 1 in 54,200
6	1	1 in 13,983,816

Table 3.1

Table 3.1 shows you the chances of guessing from zero to six numbers correctly, all calculated in the same way. To put these figures into some sort of perspective, if you bought a lottery ticket every week, then after just over a year, you might expect to have one ticket with at least three numbers correct. After about 20 years, you might have seen one ticket with at least four numbers correct. King Alfred, if he'd bought a ticket every week, might expect by now to have seen one ticket with five numbers correct. And if the first thought that entered the head of the first *Homo sapiens* was to pop down to the local newsstand and start buying a lottery ticket every week, then by now, he might have won the big prize once.

If you are ever lucky enough to win the big prize, then what you don't want to happen is what transpired in the United Kingdom on January 14, 1995, in just the ninth week of the National Lottery. The jackpot that week was over $25 million. As the six numbered balls came out of the lottery machine, the winners must have been jumping up and down on their couches and shouting for joy. But when they came to claim their prize, they each discovered that he or she had to share the jackpot with another 132 holders of winning tickets. Each winner got a paltry $195,960.

How come so many people guessed the correct combination? The reason goes back to a point I made when we were looking at the game of rock-paper-scissors: we humans are notoriously bad at choosing random numbers. Given that fourteen million people play the UK National Lottery every week, many will find themselves being drawn to very similar numbers, such as the lucky number 7 or dates of birthdays or anniversaries (which exclude the numbers 32–49). One thing in particular that typifies many people's choices is the desire to spread their numbers out evenly.

Here are the winning numbers in week 9 of the lottery:

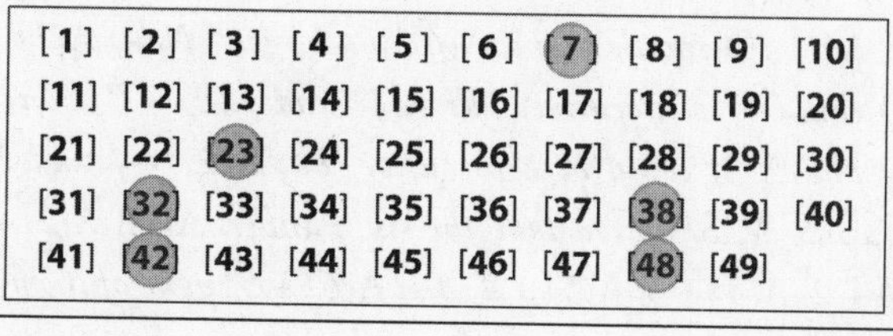

[1] [2] [3] [4] [5] [6] (7) [8] [9] [10]
[11] [12] [13] [14] [15] [16] [17] [18] [19] [20]
[21] [22] (23) [24] [25] [26] [27] [28] [29] [30]
[31] (32) [33] [34] [35] [36] [37] (38) [39] [40]
[41] (42) [43] [44] [45] [46] [47] (48) [49]

Figure 3.3

This even spacing of numbers is not particularly characteristic of randomness: numbers are as likely to be clumped together as not. Of the 13,983,816 different possible combinations of lottery tickets, 6,924,764 will have two consecutive numbers. That's 49.5 percent—very nearly half of all combinations. For example, in the previous week's draw, the numbers 21 and 22 came up. The week after, 30 and 31 appeared.

But don't get too addicted to consecutive numbers. You might think that 1, 2, 3, 4, 5, 6 is a clever choice. After all, by now you will hopefully appreciate that this combination is as likely as any other (i.e., extremely *un-*likely!). If you won the jackpot with this combination, you'd hope to walk away with the whole pot. But apparently, over ten thousand people in the United Kingdom choose this combination each week—which just goes to show how clever the UK population really is—so you'd have to share your winnings with ten thousand other clever people.

Why Numbers Like to Clump

Here's how to calculate how many lottery tickets have two consecutive numbers. Mathematicians often use a clever trick of looking at a problem the other way around, and that's what you can do here. First, you count the tickets with no consecutive numbers, then subtract the result from the total number of possible combinations to find how many combinations have consecutive numbers.

First, pick any six numbers from the numbers 1 to 44 (you'll see in a moment why you're only allowed numbers up to 44, not 49). Call your choice of numbers A(1), . . . , A(6), with A(1) the smallest and A(6) the largest. Now, A(1) and A(2) could be consecutive, but A(1) and A(2) + 1 won't be. A(2) and A(3) could be consecutive, but A(2) + 1 and A(3) + 2 won't be. So if you take the six numbers A(1), A(2) + 1, A(3) + 2, A(4) + 3, A(5) + 4, and A(6) + 5, none of them will be consecutive. (The restriction of choosing numbers up to 44 becomes clear now, because if A(6) is 44, then A(6) + 5 is 49.)

By using this trick, you can generate all the tickets with no consecutive numbers simply by picking six numbers from 1 to 44 and spreading them out by adding a little to each one. And we find that the number of tickets with no consecutive numbers is the same as the number of combinations of six numbers from 1 to 44. There are

$$\frac{44 \times 43 \times 42 \times 41 \times 40 \times 39}{6 \times 5 \times 4 \times 3 \times 2 \times 1} = 7,059,052$$

such choices. So the number of tickets with consecutive numbers is

$$13,983,816 - 7,059,052 = 6,924,764$$

HOW TO CHEAT AT POKER AND DO MAGIC USING THE MILLION-DOLLAR PRIME-NUMBER PROBLEM

Crooked gamblers and magicians don't shuffle cards the way the rest of us do. But with hours of practice, it's possible to learn how to do something called the perfect shuffle. In this shuffle, the pack is cut exactly in two and then the cards are interwoven one at a time from the two piles of cards. If you're playing poker, this shuffle is very dangerous.

Let's imagine four people sitting around the poker table: the dealer, his accomplice, and two unsuspecting gamblers who are about to be stung. The dealer puts four aces on top of the pack. After one perfect shuffle, the aces are two cards apart. After another perfect shuffle, the aces are four cards apart—perfectly placed for the dealer to deal his accomplice a hand with four aces.

The perfect shuffle comes into its own in the hands of a magician who can exploit an interesting property that it has. If you take a pack of 52 cards and do the perfect shuffle eight times, then amazingly, the cards all return to their original positions in the pack. To the onlooker, the shuffling seems to have totally randomized the pack. After all, eight shuffles done by a normal gambler is more than most do at the beginning of a game. In fact, mathematicians have proved that it takes only seven shuffles by a normal cardplayer for the pack to lose all of its original structure and become random. But the perfect shuffle is no ordinary shuffle. Think of the pack of cards as being something like an eight-sided coin, and the perfect shuffle as being the rotation of the coin by an eighth of a turn. After eight rotations, the coin comes back to its starting position.

How many times would you have to do the perfect shuffle on a pack with more than 52 cards for them to return to their original positions? If you add two jokers and do the perfect shuffle with a pack of 54 cards, it will take 52 shuffles to come full circle. But if you add another ten cards to make 64, it will now take just six perfect shuffles to return the pack to its original order. So what is the math that tells you how many times you need to perfectly shuffle a pack of $2N$ cards (it has to be an even number) to return all the cards to their starting positions?

Number the cards 0, 1, 2, and so on up to $2N - 1$, and you will see that with the perfect shuffle, the position of each card essentially doubles each time. Card 1 (which is actually the second card) becomes card 2. After another shuffle, card 2 becomes card 4, then card 8. The math is easier if we give the first card the number 0.

Where do the cards farther down the pack go? The way to work out the location of each card is to think of a clock with $2N - 1$ hours on it, so a 52-card pack is like a clock with hours from 1 to 51. If you want to know where card 32 went, then double 32, which you do by starting at hour 32 and counting 32 hours on, which gets you to 13 o'clock. To work out how many times I have to do the perfect shuffle to get all the cards back to their original position, I have to work out how many times I have to double numbers on this clock for them to return to the original position. Actually, I just have to look at the number 1 and work out how many times I have to double 1 to get back to 1. On the clock with 51 hours, here is where I get by repeatedly doubling the 1:

$$1 \to 2 \to 4 \to 8 \to 16 \to 32 \to 13 \to 26 \to 1$$

And what works for 1 will work for all the other numbers, because essentially, doing eight perfect shuffles is the same as multiplying the positions of the cards by 2^8, which is the same as multiplying by 1—that is, it leaves each card where it is.

What is the maximum number of times you would have to shuffle the pack to get back to the original order? Pierre de Fermat proved that if $2N - 1$ is prime, and you keep doubling on a $2N - 1$ clock, then after $2N - 2$ doublings, you'll definitely be back where you started. So for a 54-card pack, since $54 - 1 = 53$ is prime, 52 perfect shuffles will certainly be enough.

We need a slightly more complicated formula to calculate the maximum number of perfect shuffles if $2N - 1$ is not prime. If $2N - 1 = p \times q$, where p and q are prime, then $(p - 1) \times (q - 1)$ perfect shuffles is the maximum needed to return the pack to its original order. So for a 52-card pack, $52 - 1 = 3 \times 17$, and so $(3 - 1) \times (17 - 1) = 2 \times 16 = 32$ perfect shuffles will certainly be enough—but, in fact, you can get away with just eight perfect shuffles. (In the next chapter, I will prove Fermat's bit of magic and explain how the same bit of mathematics is at the heart of the codes used to protect secrets on the Internet.)

A mathematical question that goes back two hundred years to the work of Gauss is this: are there infinitely many numbers N with the property that a deck with $2N$ cards actually needs the full number of perfect shuffles? This question turns out to be related to the Riemann hypothesis, the million-dollar question about prime numbers that concluded chapter 1. If the primes are distributed as the Riemann hypothesis predicts they are, then there will be an infinite number of packs of cards needing the maximum number of shuffles. The Magic Circle and gamblers around the world probably aren't holding their breath waiting for the answer, but mathematicians are curious to know how primes can be related to questions of shuffling cards. It wouldn't be surprising if they were, because the primes are so fundamental to mathematics that they pop up in the strangest places.

Poker Tip

In the popular version of poker called Texas Hold'em, each player is dealt two cards facedown. The dealer then lays five cards faceup on the table. You choose the best five cards from the two in your hand and the five on the table to try to beat your opponents' hands. If you get dealt two consecutive cards (say, the 7 of clubs and 8 of spades), you might start getting excited about the possibility of a straight (five consecutive cards in any suit, like 6, 7, 8, 9, 10).

A straight is a powerful hand exactly because the chances of getting one are pretty slim, so you might think that being dealt two consecutive cards is worth a big stake because you're on your way to a straight. Now, this is when you need to remember the lottery tip. Two consecutive numbers come up very often in the lottery, and the same is true of poker. Did you know that over 15 percent of starting hands dealt in Texas Hold'em have two consecutive cards? But slightly less than a third of these will go on to complete a straight by the time the dealer has dealt the five cards on the table.

THE MATH OF THE CASINO: DOUBLE OR BUST?

You're in the casino at the roulette wheel, and you have 20 chips. You've decided to try to double your money before you leave. Putting a chip on red or black will double your money if you choose correctly, so what is the best strategy—putting all your money in one go on red, or putting one chip on at a time until you've either lost all your money or you've got your 40 chips?

To analyze this problem, what you have to realize is that every time you place a bet, you are essentially paying the casino a small amount to play, once you average out over all your wins and losses. If you put your money on black 17 and it comes up, then the casino gives you your chip back along with 35 more. If there were 36 numbers on the roulette wheel, this would be a fair game, since on average, black 17 would come up 1 in 36 times. So if you had 36 chips and kept on betting on 17, then in 36 spins of the wheel, on average 35 of them would lose and one would win, leaving you with the 36 chips you started with. But on the European roulette wheel, there are actually 37 numbers you can bet on (1 to 36 together with 0, which is neither red nor black), but the house pays out as if there were only 36 numbers.

Because there are 37 numbers, every time you bet $1, the house is essentially making $\frac{1}{37} \times \$1$, which is about 2.7 cents. Every now and again, the casino might have to make a big payout to one individual, but in the long run, it knows that, thanks to the laws of probability, it makes money. In fact, the house odds in the United States are even worse for the gambler because casinos use roulette wheels with 38 numbers: 1 to 36, plus 0 and 00.

We've seen that betting on a single number costs you, in the long run, 2.7 cents per bet. But you don't have to bet on a single number: you can bet, for example, on the number being odd or even, or being in the range from 1 to 12. The odds are calculated in the same way, such that whatever sort of bet you make essentially costs you 2.7 cents per bet.

So what should you do to give yourself the best chance of doubling your money? First, since you pay every time you play, the best strategy is to play as few times as possible. There is an 18/37 chance, just under 50 percent, that red will come up and you'll walk away with double your money, so though it will be a short visit to the casino, the best strategy for doubling your money is to put it all on red in one go. The likelihood of doubling your money by putting on a chip at a time comes out at

$$\frac{1 - \left(\frac{19}{18}\right)^{20}}{1 - \left(\frac{19}{18}\right)^{40}}$$

which is a 25.3 percent chance. You halve your chances of achieving your goal if you bet one chip at a time.

But where is the best place to bet in roulette? If you put your money on red and 0 comes up, some casinos will apply a rule called *en prison* and pay you back half your bet. This actually means that the house odds are a little less on this bet—it's cheaper to play here than anywhere else on the roulette wheel. In the long run, it costs you (probability of losing) × bet − (probability of winning) × payout = 1.35 cents, as opposed to the 2.7 cents it costs to play anywhere else on the table:

$$= \frac{18}{37} \times \$1 + \frac{1}{37} \times \$0.50 - \frac{18}{37} \times \$1$$

So if the casino plays *en prison,* in the long run, it is half the price to bet on red/black than to make other types of bets.

Instead of getting half your bet back, there is another option that the casino can offer: you can choose to have your bet go *en prison.* The dealer puts an *en prison* chip on the bet, and if red comes up next, then you get a reprieve and the casino gives you your bet back (but without any winnings); otherwise, you lose your bet. Because there is an $18/37$ chance of you then getting all your money back (just under 50 percent), you are better off taking half the money when you have the chance to rather than putting your bet in prison and hoping for red to come up.

So, the odds are apparently stacked against you. But is there any mathematical way you can beat the casino? Here's one idea, called a martingale. Start by putting one chip on red. If red comes up, you get your chip back plus another chip. If it isn't red, then on the next round, bet two chips on red. If it comes up red, you get your chips back plus two more. You lost one chip with the first bet, so you're now one chip up. If red fails to come up the second time, bet four chips next time. If red comes up then, you get four chips on top of your bet. But you've already lost one chip on the first bet and two on the second, so that leaves you . . . one chip up.

The way to play this system is to keep doubling your bet each time until red eventually comes up. Your total winnings are always one chip, because if red comes up on round N, then you win 2^N chips (the amount you bet). But in the previous $N-1$ rounds, you'll have lost $L = 1 + 2 + 4 + 8 + \ldots + 2^{N-1}$ chips. Here's a clever way to calculate how much this loss L is. L is certainly the same as $2L - L$. So how much is $2L$?

$$2L = 2 \times (1 + 2 + 4 + 8 + \ldots + 2^{N-1}) = 2 + 4 + 8 + 16 + \ldots + 2^{N-1} + 2^N$$

Now take away

$$L = 1 + 2 + 4 + 8 + \ldots + 2^{N-1}$$

This gives

$$L = 2L - L = (2 + 4 + 8 + 16 + \ldots + 2^{N-1} + 2^N) -$$
$$(1 + 2 + 4 + 8 + \ldots + 2^{N-1}) = 2^N - 1$$

All the numbers in the first set of parentheses except for 2^N also appear in the second set of parentheses, which is why they all disappear in this calculation! (We've met this calculation before, when we were piling grains of rice on a chessboard in our search for prime numbers in chapter 1.) So you win 2^N, but you've lost $2^N - 1$. Your net gain is one chip.

It's not a lot, but the system appears to guarantee you a win eventually—after all, at some point, red is surely going to come up . . . isn't it? So why aren't gamblers cashing in at the casinos with this strategy? One problem is that you would need an infinite amount of resources to guarantee a win, since there is a theoretical possibility of a run of blacks all night long. And even if you had a huge pile of chips, repeatedly doubling your bet can very quickly exhaust your supply (as with those rice grains). On top of that, most casinos have a maximum limit on bets precisely to stop players from exploiting this strategy. For example, with a maximum bet of one thousand chips, your strategy is going to fail after nine rounds because on the tenth round, you are going to want to bet $2^{10} = 1,024$—already more than the maximum bet.

Even with a maximum bet in place, the gambler's fallacy is to believe that if there have been eight blacks in a row, the probability must be really high of seeing red come up next. Of course, the chance of seeing eight blacks in a row is incredibly small—1 in 256, in fact. But that won't increase the chances of getting red next: that's still fifty-fifty. Like the tossed coin, the roulette wheel has no memory.

If you want to play roulette, then bear in mind what the mathematics of probability says: in the long term, the house always wins—although as we shall see in chapter 5, there might be a way to use some other mathematics to help you make your millions. If you don't like poker or roulette, then the craps table might be for you. As we will now see, playing with dice has a very long history.

HOW MANY FACES DID THE FIRST DICE HAVE?

Many of the games we play depend on chance. Monopoly, backgammon, snakes and ladders, and many others rely on the throw of dice to determine how many steps you move your counter. The very first dice were thrown by the ancient Babylonians and Egyptians, who used knucklebones—the "ankle" bones of animals such as sheep—as dice.

The bones would naturally land on one of four sides, but ancient gamers soon realized that given the uneven nature of bones, some sides were favored over others, so they started to craft them to make the game fairer. As soon as they started doing this, they found themselves exploring the variety of three-dimensional shapes for which each face is equally likely to be the one the shape lands on.

Because the first dice evolved from knucklebones, it's not too surprising that some of the first symmetrical dice to be made were in the shape of the tetrahedron, with four triangular faces. One of the earliest board games we know of uses these pyramid-shaped dice.

Called the Royal Game of Ur, several versions of the board and its tetrahedral dice were discovered in the 1920s by British archaeologist Sir Leonard Woolley while he was excavating tombs in the ancient Sumerian city of Ur, in what is today southern Iraq. The tombs date back to 2600 BC, and the boards would have been placed in the tombs to keep their occupants amused in the afterlife. The finest example is on display in the British Museum in London and consists of 20 squares that the opponents must race around, depending on the throw of the tetrahedral dice.

Figure 3.4   Tetrahedral dice from the Royal Game of Ur.

The rules of the game didn't come to light until the early 1980s, when Irving Finkel at the British Museum stumbled across a cuneiform tablet from 177 BC in the museum's archives, which had a picture of the game engraved on the back. The game is an early forerunner of backgammon, and each player has a certain number of pieces that he or she must move around the board. But it is the dice associated with the game that are most interesting from a mathematical point of view.

One problem with tetrahedral dice made from four triangles is that, unlike the cube-shaped dice we are all familiar with, tetrahedrons land with one of the points pointing in the air, not a face. To deal with this, two of the four corners of each die would be marked with white dots. Players would throw a number of pyramids, and their score would correspond to the number of dots uppermost. Throwing these dice is mathematically equivalent to tossing a number of coins and counting the number of heads.

The Royal Game of Ur depends heavily on the random outcome of the throw of the dice. In contrast, backgammon, its successor, provides players with more opportunity to show off their skills and strategy rather than relying solely on the luck of the dice. But the game has not died out completely: it recently came to light that Jews from Cochin in Southern India are still playing a version of the Royal Game of Ur—five thousand years after it was played in ancient Sumer.

DID DUNGEONS AND DRAGONS DISCOVER ALL THE DICE?

One of the novelties of Dungeons and Dragons, the fantasy role-playing game from the 1970s, was its intriguing array of dice. But did the inventors of the game discover all the dice possible? When we look at what shapes make good dice, we come to a question we asked in chapter 2. If all the faces of the dice are the same symmetrical shape and these faces are arranged such that all the vertices and edges look the same, then there are five such dice: the tetrahedron, the cube, the octahedron, the dodecahedron, and the icosahedron—the Platonic solids (page 61). You will find all these dice in the Dungeons and Dragons box (and on a PDF you can download from the Number Mysteries website), but many of them have a much older heritage.

For example, a 20-faced die made of glass dating back to Roman times was sold by Christie's in 2003. Its faces are carved with strange symbols, suggesting that it might have been used in fortune telling rather than in a game. The icosahedron is at the heart of one of today's most fashionable fortune-telling devices: the Magic 8 Ball. Floating in fluid inside the ball is an icosahedron with answers to your problems written on the faces. Ask a question, shake the ball, and the icosahedron floats to the top revealing the answer on one of its faces. These range from "Without a doubt" to "Don't count on it."

If you just want a fair die, you don't have to be so strict about the arrangement of faces. For example, Dungeons and Dragons used a die made by fusing two pentagonal-based pyramids together at their bases. This die has a one-in-ten chance of landing on any of the ten triangular faces. It's not a Platonic solid because the vertex at the tip of each pyramid is distinguishable from all the other vertices: five triangles meet there, while each of the vertices where the two bases meet is a conjunction of four triangles. But it is still a fair die: it is equally likely to land on any one of the ten faces.

Mathematicians have been investigating what other shapes make fair dice. It was proved relatively recently that if the dice still have some symmetry to them, there are another 20 to add to the five Platonic dice, together with five infinite families that make fair dice.

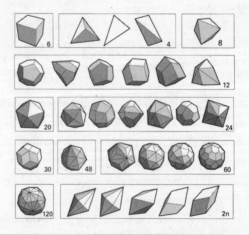

Figure 3.5 Symmetrical shapes that make good dice and the number of faces on each shape.

Of the extra 20, 13 are related to the shapes that make great soccer balls—the Archimedean solids of chapter 2, which have faces that are all symmetrical but need not be all the same shape. These shapes may make great soccer balls, but they aren't quite right for dice. The classic soccer ball has 32 faces made up of 12 pentagons and 20 hexagons. Couldn't we make a fair die by just writing the numbers 1 to 32 on the faces? The problem is that each pentagon has roughly a 1.98 percent chance of being chosen, while each hexagon has a 3.81 percent chance, so this wouldn't be a fair die. It was only in the last decade that mathematicians produced a precise formula for the probability that the soccer-ball die would land with a pentagon uppermost. An impressive bit of geometry produced the following scary answer:

$$12 \times \frac{-3 + 30r\left[1 - \left(\frac{2}{\pi}\right)\sin^{-1}\left(\frac{1}{2r\sqrt{3}}\right)\right]}{-116 + 360r}$$

where $r = \frac{1}{2}[2 + \sin^2(\pi/5)]^{-1/2}$.

The Archimedean solids themselves are not fair dice, but they can be used to build different shapes that give a whole new selection of dice for gamers to use. The key is to realize that although the faces might vary around an Archimedean solid, the vertices are all the same, and the trick is to use an idea called duality, which changes the points into faces and vice versa. To see what shape the face should be, you need to think of a sheet of card stock being placed on each point and then look at how all these cards intersect or cut into one another. Each card needs to be angled so that it is perpendicular to the line running from the center of the shape to the vertex. For example, if you replace the vertices of a dodecahedron with faces, you get the icosahedron:

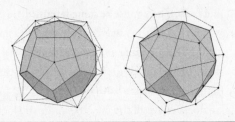

Figure 3.6

By playing this trick with the Archimedean solids, the procedure produces 13 new dice. The classic soccer ball has 60 vertices, and the die that emerges from it when we replace each vertex with a new face is made up of 60 triangles, which are not equilateral but isosceles (i.e., only two of the three sides are equal). Although this dual of the classic soccer ball is not a Platonic solid, it is still a shape in which each of the faces has a 1 in 60 chance of coming up, so it makes a fair die for gamers to play with. Its technical name is the pentakis dodecahedron:

Figure 3.7

Each Archimedean solid can be used to create a new die like this. Perhaps the most impressive is the hexakis icosahedron. Amazingly, even with 120 irregular right-angled triangular faces, this shape gives another fair die.

The infinite number of families of dice comes from generalizing the idea of sticking two pyramids together where the base can have any number of edges. Although mathematicians have sorted out the range of fair dice that have symmetry, there is still a mystery about more irregular shapes that make fair dice. For example, if I take an octahedron and cut a little bit off one vertex and the opposite vertex, two new faces appear. If I throw this shape, it is unlikely to land on one of these new faces, but if I cut larger chunks off, these two faces will be more likely to appear uppermost than the eight remaining faces. There must be some intermediate point at which I can cut these two corners off such that the two new faces and the original eight faces are equally likely, creating a fair die with ten faces.

The shape doesn't have any of the nice symmetry of the new dice we made from the Archimedean soccer balls, but it too would make a fair die. As proof that math doesn't have all the answers, we are still looking for a way to classify all the shapes that can be concocted like this and make fair dice.

HOW CAN MATH HELP YOU WIN
AT MONOPOLY?

Monopoly appears to be a pretty random game. You throw two dice and speed around the board in your car or strut along in your top hat, buying property here, building hotels there. Every now and again you might come second in a beauty contest thanks to a Community Chest card or have to cough up $20 for drunk driving. Each time you pass GO, you collect another $200. How on earth can math give you an edge in this game?

Over the course of the game, which is the most visited square on any Monopoly board? Is it the GO square where you start, the Free Parking diagonally opposite, or perhaps Virginia Avenue or the Boardwalk? The answer is in fact the Jail square. Why? Well, you could just throw the dice and find yourself just visiting, or you might find that the dice takes you to the square diagonally opposite, where a policeman tells you to go to jail. You might even be unlucky and pick up one of the Chance or Community Chest cards that send you straight to jail. And if that wasn't enough ways to send you down, if you throw a double, you get to go again, but if you throw three doubles in a row, then rather than being rewarded for your impressive feat of dice rolling, that too is punished with a three-turn sentence.

As a result, on average, players find themselves visiting the Jail square about three times more often than most other squares on the board. That isn't much help to us at the moment, because you can't buy the jail. But here is where the math comes to the fore: where are players most likely to land after being in jail? The answer depends on the most likely throw of the dice when they leave that square.

Each die can land equally on one of the six faces. With two dice, that gives $6 \times 6 = 36$ different possible throws, each equally likely. But when you analyze those possibilities, you find that a score of 2 or 12 is very unlikely, because there's only one way to make either of these combinations, whereas there are six ways to make a total score of 7.

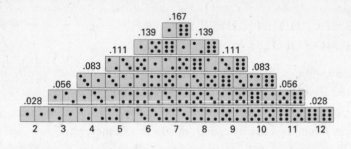

Figure 3.8

So, there's a 6 in 36 or 1 in 6 chance of getting a 7, and scores of 6 and 8 are the next most likely. A throw of 7 from jail gets you to a Community Chest square, which you can't buy, but the two orange properties on either side (Tennessee Avenue and St. James Place) are the next most likely stops.

If you are lucky enough to land in the orange region of properties, these are the ones to buy and stack with hotels while you sit back and collect the rent that all your opponents will have to pay you as the dice take them out of jail and straight to your lair.

THE NUMBER MYSTERIES GAME SHOW

This is a game for two players. Take 20 envelopes and number them from 1 to 20. Player 1 writes down 20 different sums of money on pieces of paper and puts one in each envelope. Player 2 then chooses an envelope and is offered the sum of money inside. He can accept the money or choose a different envelope. If he chooses a different envelope, he can't go back and claim a previous prize.

Player 2 continues opening envelopes until he is happy with the prize he has. Player 1 then reveals all the prizes. Player 2 scores 20 points if he claimed the top prize on offer. He scores 19 points if he got the second best prize, and so on.

All the envelopes are now emptied, and player 2 writes down 20 different sums of money on pieces of paper and puts one in each envelope. Player 1 must now try to get the best prize she can. Once she settles on an envelope, she scores points in the same way player 2 did. The winner is whoever has the highest score. This doesn't mean the highest amount of money, but the highest number of points.

The intriguing aspect of this game is that you don't know what the range of prizes is: the top prize might be $1, or it might be $1,000,000. The question is whether there's a mathematical strategy that will help you to increase your chances of winning. Well, there is. It's in a secret formula that depends on e—not the psychedelic kind, but the mathematical kind. The number e = 2.71828 . . . is probably one of the most famous numbers in the whole of mathematics, second only to the enigmatic π, and crops up wherever the concept of growth is important. For example, it is intimately related to the way the interest accumulates in your bank account.

Imagine that you have $1 to invest and are looking to see what different interest-rate packages the banks are offering. One will pay 100 percent interest after one year, which would increase your investment to $2. Not bad, but the next bank offers to pay 50 percent interest every half-year. After six months, that would give you $1.50, and after a year, $1.50 + $0.75 = $2.25—a better deal than the first bank. A third bank is offering 33.3 percent added every third of a year, which comes out at $(1.333)^3 = $2.37 after 12 months. As you slice the year into smaller and smaller chunks, this compounding of the interest works to your advantage.

By now, the mathematician in you will hopefully have realized that the bank you really want is the Bank of Infinity, which divides the year into infinitely small units of time, because this will give you the maximum balance you can achieve. Although the balance increases the more you divide the year, it doesn't become infinite but tends instead toward this magic number, e = 2.71828 . . . . Like π, e has an infinite decimal expansion (indicated by the " . . ."), which never repeats itself. It turns out to be the key to helping you win the Number Mysteries game show.

The mathematical analysis of this game implies that you should first calculate $\frac{1}{e}$, which is about 0.37. You should start by opening 37 percent of the envelopes, or about seven of them. Continue to open envelopes, but stop at the one whose contents beats all the envelopes you've opened so far. The math implies that one out of three times this will ensure that you end up with the top prize on offer. This strategy isn't just useful in playing the Number Mysteries game show. In fact, many decisions we make in our lives can be reached by adopting this tactic.

Remember the first boyfriend or girlfriend you had? You probably thought he or she was amazing. Perhaps you dreamed romantically of spending your life together, but then had that nagging feeling that maybe you could do better. The problem is that if you dump your current partner, there's generally no way back, so at what point should you just cut your losses and settle for what you've got? House hunting is another classic example. How many times do you see a fantastic house on your first visit, but then feel you need to see more before you commit yourself, only to risk losing the first great house?

Amazingly, the same math that helps you win the Number Mysteries game show can give you the best chances of landing the best partner or the best house. Let's say you start dating at the age of 16 and decide that you'll aim to have found the love of your life by the time you reach 50. And let's assume that you get through partners at a constant regular rate. The math says that you should survey the scene for the first 37 percent of the time you've set yourself, which takes you to about the age of 28. Then you must choose the next partner who's better than all the people you've dated up to that point. For one in three people, this will ensure that they end up with the best partner possible. Just be sure not to reveal your method to the love of your life!

HOW TO WIN AT CHOCOLATE ROULETTE

Even if you know your math, games like Monopoly or the Number Mysteries game show still rely on chance. Here's a simple game for two players, which illustrates how math can guarantee you a win every time. Take 13 bars of chocolate and a red-hot chili and put them in a pile on the table. Each player, in turn, takes one, two, or three items from the pile. The aim is to force your opponent to take the chili.

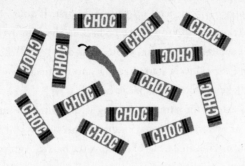

Figure 3.9   *Chocolate roulette.*

Provided you go first, there is a strategy that will always leave your opponent with the chili. However many bars of chocolate your opponent takes, you always remove the number of bars that makes the total taken during the round add up to four. For example, if your opponent takes three bars of chocolate, you take one, making four bars in total. If your opponent takes two, you also take two.

The trick is to arrange the bars of chocolate in rows of four (do this in your head; otherwise, you give the game away). There are 13 bars to start with, so that's three piles of four with one bar remaining (plus the chili, of course). Your opening move, then, is to take the one bar of chocolate leftover. After that, you carry on as just described: in response to your opponent's move, you take a number that adds to four. In this way, the combination of your opponent's move and your move removes one of the piles of four chocolates each time. After three rounds, your opponent is left with just the chili on the table.

Figure 3.10   *How to arrange the chocolates to guarantee a win.*

The strategy does depend on you going first. If your opponent goes first, it only takes one slip for you to be back in the winning position. For example, if your opponent takes more than one bar of chocolate as his opening move, he's already started eating into the first pile of four chocolates, so you take the rest of the pile as before.

You can extend the game by starting with a different number of bars of chocolate or by varying the maximum number of bars you are allowed to take in one turn. The same math of dividing the bars into groups will enable you to concoct a winning strategy.

There is another variant of this game, called nim, which uses a slightly more sophisticated mathematical analysis to guarantee you your win. There are four piles this time. One pile has five bars of chocolate, the second pile has four, the third pile has three, and the final pile consists of just the chili. This time, you are allowed to take as many bars of chocolate as you want, but they can only be taken from one pile. For example, you could take all five bars of chocolate from the first pile, or just one bar of chocolate from the third pile. Again, you lose if your only choice is to take the chili.

The way to win this game is to know how to write numbers in binary rather than decimal form. We count in 10s because we've got ten fingers. Once you've counted up to 9, you start a new column and write 10 to indicate one lot of ten and no units. But computers like to count in 2s, or what we call binary. Each digit represents a power of 2 rather than a power of 10. For example, 101 represents one lot of $2^2 = 4$, no 2s, and one unit. So 101 is the number $4 + 1 = 5$ in binary. The table here shows the first few numbers written in binary.

Decimal	Binary
0	0
1	1
2	10
3	11
4	100
5	101
6	110
7	111
8	1000
9	1001

Table 3.2

To win at nim, you need to convert the number of bars of chocolate in each pile into binary. The first pile has 101 bars, the second 100 bars, the third 11 bars. Writing this last number as 011 and putting the three numbers on top of each other give us this:

$$101$$
$$100$$
$$011$$

Notice that the first column has an even number of 1s; the second, an odd number of 1s; and the third column, an even number of 1s. The winning move each time is to remove bars of chocolate from one pile in such a way that each column ends up with an even number of 1s. So in this case, remove two bars from the third pile of three to reduce the number of bars to 001.

Why will this help you win? Well, at every turn, your opponent will be forced to leave at least one of the columns with an odd number of 1s. Your next move is to take bars of chocolate to make them all even again. Because the number of bars of chocolate is constantly going down, at some point, one of you will remove bars of chocolate so that the piles have 000, 000, and 000 in them. Who does that? Your opponent always leaves an odd number of 1s in at least one of the piles, so it must be you that makes this move. Your opponent gets left with the chili.

This strategy will work however many bars of chocolate you put in each pile. You can even increase the number of piles.

WHY ARE MAGIC SQUARES THE KEY TO EASING CHILDBIRTH, PREVENTING FLOODS, AND WINNING GAMES?

Lateral thinking is a handy talent when it comes to doing math. By looking at things from a different angle, the answer to a difficult conundrum can suddenly become transparently obvious. The skill lies in finding the right way to look at the problem. To illustrate this, here's a game that, at first glance, is tricky to keep track of but becomes much easier when we come at it from a different direction. To play the game, you can visit the Number Mysteries website to download and cut out the props you'll need.

Each contestant has an empty cake stand on which can fit 15 slices of cake. The object of the game is to be the first to fill the cake stand with exactly three chunks of cake chosen from nine chunks of different sizes, the smallest consisting of just one slice and the largest consisting of nine slices. Contestants take turns choosing one of the chunks.

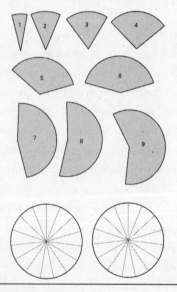

Figure 3.11   Choose three chunks of cake to fill your cake stand before your opponent fills his.

The aim is to get three numbers from 1 to 9 that add up to 15 while at the same time keeping track of what your opponent is doing so you can defeat his attempts. So if your opponent has chosen chunks with three and eight slices, then you need to stop him from making 15 by taking the chunk with four slices. If the chunk you wanted has been taken, you'll have to find a different way to get to 15 by using the chunks you've already chosen and the remaining ones. But you must always use precisely three chunks to fill the stand—filling the stand with two chunks of nine and six slices is not a winning move, nor is filling it with four chunks of one, two, four, and eight slices.

Once you start playing, it soon gets quite difficult to keep track of all the different ways that you and your opponent might fill the cake stand. But the game becomes much easier once you realize that what you're playing is actually another classic game in disguise: tic-tac-toe. Instead of the classic 3 × 3 grid in which you place Os and Xs, trying to get three in a row before your opponent does, this game is played out on a magic square:

2	9	4
7	5	3
6	1	8

Table 3.3

The most basic magic square is a way of arranging the numbers from 1 to 9 in a 3 × 3 grid so that the numbers in the columns, rows, and diagonals all add up to 15. This arrangement provides all the possible ways to get 15 by adding three different numbers chosen from 1 to 9. By playing the cake game as tic-tac-toe on this magic square, anyone who gets three in a row will have gotten three numbers that add up to 15 before his or her opponent has.

According to one legend, the first magic square appeared in 2000 BC inscribed on the back of a turtle that crawled out of the River Lo in China. The river had badly flooded, and the emperor Yu ordered a number of sacrifices to appease the river god. In response, the river god sent forth the turtle, whose pattern of numbers was meant to assist the emperor in controlling the river. Once this arrangement of numbers had been discovered, Chinese mathematicians started trying to construct bigger squares that worked the same way. These squares were believed to have great magical properties and became widely used for divination. The Chinese mathematicians' most impressive achievement was a 9 × 9 magic square.

There is evidence that the squares were taken to India by Chinese traders who dealt not only in spices but also in mathematical ideas. The way the numbers weaved in and out of the squares resonated strongly with Hindu beliefs of rebirth, and in India, these squares were used for anything from specifying perfume recipes to an aid for childbirth. Magic squares were also popular in medieval Islamic culture. Their much more systematic approach to mathematics led to clever ways of generating magic squares, culminating in the thirteenth-century discovery of an impressive 15 × 15 magic square.

One of the earliest outings of magic squares in Europe is the 4 x 4 magic square that appears in Albrecht Dürer's engraving *Melancholia*. Here, the numbers from 1 to 16 are arranged so that the rows, columns, and diagonals all add up to 34. As well as that, each of the four quadrants—the four 2 × 2 squares into which the big square can be split—and the 2 × 2 square at the center also sum to 34. Dürer even arranged the two numbers at the middle of the bottom row to give the year he made the engraving: 1514.

Figure 3.12   Albrecht Dürer's magic square.

Magic squares of different sizes were traditionally associated with planets in the solar system. The classic 3 × 3 square was associated with Saturn, the 4 × 4 square in *Melancholia* is Jupiter's, while the largest—a 9 × 9 square—was assigned to the moon. One suggestion for Dürer's use of the square is that it reflected the mystical belief that Jupiter's joyfulness could counteract the sense of melancholy that pervades the engraving.

Another famous magic square can be found at the entrance to the flamboyant Sagrada Família, the still unfinished cathedral in Barcelona designed by Antoni Gaudí. The magic number for this 4 × 4 square is 33, the age Christ was when he was crucified. This square isn't quite as satisfying as Dürer's square because the numbers 14 and 10 appear twice, at the expense of 12 and 16.

Magic squares are something of a mathematical curiosity, but there is a problem about them that mathematicians have been unable to unravel. There is essentially only one 3 × 3 magic square. (The qualification "essentially" means that what you get by rotating or reflecting a magic square doesn't count as a different one.) In 1693, the Frenchman Bernard Frénicle de Bessy listed all 880 possible 4 × 4 magic squares, and in 1973, Richard Schroeppel used a computer program to calculate that there are 275,305,224 5 × 5 magic squares. Beyond that, we only have estimates for the number of possible magic squares of 6 × 6 and bigger. Mathematicians are still looking for a formula that will give the exact numbers.

WHO INVENTED SUDOKU?

The spirit of sudoku can be found in a puzzle that grew out of mathematicians' fascination with magic squares. Take the court cards (kings, queens, jacks) and aces from a standard pack of cards and try to arrange them in a 4 × 4 grid so that no row or column has a card of the same suit or rank. This problem was first posed in 1694 by the French mathematician Jacques Ozanam, who might be regarded as the man who invented sudoku.

One mathematician who certainly caught the bug was Leonhard Euler. In 1779, a few years before he died, Euler came up with a different version of the problem. Take six regiments, with six soldiers in each regiment. Each regiment has a different colored uniform: they might be red, blue, yellow, green, orange, and purple. The soldiers in each regiment have different ranks—say, a colonel, a major, a captain, a lieutenant, a corporal, and a private. The problem is to arrange the soldiers on a 6 × 6 grid so that in any column (or row) you don't see a soldier of the same rank or in the same regiment in that column (or row). Euler asked the question for a 6 × 6 grid because he believed that it was impossible to arrange the 36 soldiers satisfactorily. It was not until 1901 that the French amateur mathematician Gaston Tarry proved Euler correct.

Euler also believed that the puzzle was impossible to solve for a 10 × 10 grid, a 14 × 14 grid, an 18 × 18 grid, and so on, adding four each time. Not so, it turned out. In 1960, with the aid of a computer, three mathematicians showed that it was in fact possible to arrange ten different ranks of soldiers from ten regiments in a 10 × 10 grid in the way Euler thought was impossible. They went on to disprove Euler's hunch completely, showing that the 6 × 6 grid is the only one in which his arrangement is impossible.

If you want to try the 5 × 5 version of Euler's puzzle, then download the appropriate file at the Number Mysteries website, cut out the five ranks in five regiments and see if you can arrange them in a 5 × 5 grid so that in any column or row you don't see a soldier of the same rank or in the same regiment. These magic squares are sometimes called Graeco-Latin squares. Take the first n letters of the Latin and Greek alphabets and write down all the n × n different pairs of Latin and Greek letters. Now arrange these pairs on an n × n grid so that no row or column contains the same Latin or Greek letter.

Living by the Square

One of the 10 × 10 Graeco-Latin squares was used by the French novelist Georges Perec to structure his 1978 novel Life: A User's Manual. *The book has 99 chapters, each corresponding to a room in a Parisian apartment block that has ten floors and ten rooms on each floor (one room, the 66th, doesn't get visited). Each room corresponds to a position in a 10 × 10 Graeco-Latin square. But in Perec's square, instead of ten Greek and ten Latin letters, he uses, for example, 20 authors divided into two lists of ten. When he wrote the chapter for a particular room, he looked to see which two authors were assigned to that room and made sure that he quoted passages from these authors during the course of the chapter. For chapter 50, for example, Perec's Graeco-Latin square told him to quote Gustave Flaubert and Italo Calvino. But it isn't only authors that figure in this scheme. Perec used a total of 21 different Graeco-Latin squares, each one filled with two sets of ten items ranging from furniture, artistic style, and a period in history through to the body positions adopted by the occupants of the rooms.*

Sudoku works slightly differently than Euler's soldiers puzzle. In the classic form, you have to arrange nine lots of the numbers from 1 to 9 on a 9 × 9 grid so that no row, column, or 3 × 3 quadrant contains a number twice. A few of the numbers are already placed on the grid, and you have to fill in the rest. Don't believe those who say that no math is needed to do these puzzles. What they mean is that there's no arithmetic involved—sudoku is essentially a logic puzzle. The kind of logical reasoning that leads you to decide that a 3 must go in the lower right-hand corner is precisely what mathematics is about.

There are some interesting mathematical questions relating to sudoku. One is this: how many different ways are there of arranging the numbers in the 9 × 9 grid to satisfy the sudoku rules? (Again, we mean "essentially" different: we regard two arrangements as the same if there is some simple symmetry, like swapping rows around, which changes one into another.) The answer was calculated in 2006 by Ed Russell and Frazer Jarvis to be 5,472,730,538—enough to keep the newspapers going for a while yet.

Another mathematical problem that arises from these puzzles has not been completely resolved. What is the minimum number of squares that need to have a number in them to start with for there to be just one unique way to fill in the other squares? Clearly, if you have too few—say, three—numbers in the grid, there will be many ways to complete the grid because there just isn't enough information to force a unique solution. It is believed that you need at least 17 numbers to ensure that there is only one way to complete a sudoku puzzle. These questions are more than just recreational puzzles. The mathematics underlying sudoku has important implications for the error-correcting codes that we'll meet in the next chapter.

HOW CAN MATH HELP YOU GET FROM A TO B?

During the eighteenth century, residents of the Prussian city of Königsberg were stumped by a problem about how to navigate their city. The River Pregel has two branches that run around the island at the heart of the city before emerging to the west and flowing into the Baltic. At the time, there were seven bridges spanning the Pregel, and it became a Sunday-afternoon pastime among the residents of the city to see whether they could find a way to cross all the bridges once and once only. But however hard they tried, they always found that there was one bridge they couldn't get to. Was it really a mission impossible, or was there some route the residents hadn't yet taken by which they could cross all seven bridges?

The problem was finally resolved by Leonhard Euler, the Swiss mathematician who'd posed the problem about Graeco-Latin squares, when he was teaching at the academy in St. Petersburg, some five hundred miles northeast of Königsberg. Euler made an important conceptual leap. He realized that the actual physical dimensions of the town were irrelevant: what mattered was how the bridges were connected together (the same principle applies to the topological map of the London Underground or the New York Subway). The four regions of land connected by the bridges of Königsberg can each be condensed to a point, leaving the bridges as lines connecting the points. This gives a map of the bridges of Königsberg that's like a much simpler London Underground map:

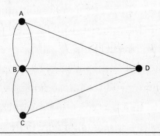

Figure 3.13

The problem of whether there is a journey around all the bridges then boils down to asking whether it's possible to trace over this map without taking your pen off the paper or running over any line twice. From Euler's new mathematical perspective, he could see that it was in fact impossible to cross all seven bridges once and only once.

So why is it impossible? As you draw the map, each point visited in the middle of a journey must have one line into it and one line out. If you visit that point again, it will be by crossing a new "bridge" into it and a new "bridge" out of it. So there must be an even number of lines attached to each point, except at the beginning and the end of the journey.

If we look at the plan of the seven bridges of Königsberg, we can see that at each of the four points, an odd number of bridges meet—and that tells us that there is no route around the city that crosses each of the bridges only once. Euler took his analysis further. If the map has precisely two points with an odd number of lines coming out, then it is possible to trace over it without taking your pen off the page and without running over any line twice. To do this, you have to start at one of the points with an odd number of lines coming out and aim to finish at the other point where an odd number of lines meet.

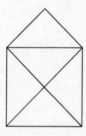

Figure 3.14   Euler's theorem implies that it's possible to draw this map without taking your pen off the paper and going over a line twice.

There is a second sort of map on which it's possible to follow what mathematicians now call an Eulerian path: one in which every point has an even number of lines running out of it. On a map like this, you can start anywhere you like, because the path must start and end at the same point so that it traces a closed loop. Even though you might have difficulty actually identifying the pathway, Euler's theorem tells you that, as long as the map is one of the two types I've described, there must be an Eulerian path. This is the power of mathematics: it can quite often tell you that something must exist without your having to construct it.

To prove that such a path exists, we make use of a classic weapon in the mathematician's arsenal: induction. It's like what I do to get over my fear of heights when I'm climbing high ladders or rappelling down waterfalls: take one small step at a time.

Start by imagining that you know how to draw all the maps with a certain number of edges without taking your pen off the page. But now you're faced with a map that has one more edge than the ones you've met so far. How do you know that you can still draw this new map?

Let's say it's a map that has two points with an odd number of edges coming out of them, and let's call these points A and B. The trick is to remove one of the edges from one of the points with an odd number of edges. So let's remove an edge going from B to another point, C. This new map with one edge removed still has only two points with an odd number of edges coming out of it: A and C. (B now has an even number because we've just removed one line; C now has an odd number because we've removed the line joining it to B.) This new map is now small enough to be drawn, with a path starting at A and finishing at C. The bigger map is also now simple to draw: just join C to B, adding in the edge we removed earlier. Bingo!

There are a few variations that we need to analyze. For example, what if there is only one line from B that joins it to A so that A and C are the same point? But we can see that at the heart of Euler's proof is this beautiful idea of building up step by step why an Eulerian path must be possible. Just as with steadily climbing a ladder, I can use this trick to find my way around however large a map I might encounter.

To see the power of Euler's theorem, challenge a friend to draw as complicated a map as she wants. Then, by simply counting the number of points where an odd number of lines meet, thanks to Euler's theorem you can tell instantly whether the map can be drawn without taking your pen off the page and without running over a line twice.

I recently went on a pilgrimage to Königsberg, which was renamed Kaliningrad after the Second World War. The city is unrecognizable from Euler's day—it was devastated by Allied bombing. But three of the prewar bridges were still in place: the Timber Bridge (Holzbrücke), the Honey Bridge (Honigbrücke), and the High Bridge (Hühe Brücke). Two of the bridges had disappeared completely: the Offal Bridge (Küttelbrücke) and the Blacksmith's Bridge (Schmiedebrücke). The remaining bridges—the Green Bridge (Grüne Brücke) and the Merchant's Bridge (Krämerbrücke)— although destroyed during the war, had been rebuilt to carry a two-lane highway through the city.

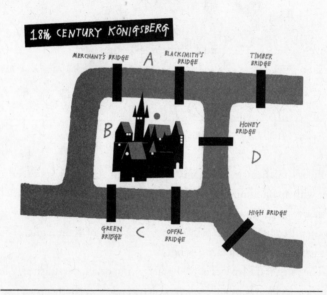

Figure 3.15   The bridges of Königsberg in the eighteenth century.

A new railway bridge, which pedestrians can also use, now joins the two banks of the Pregel out to the west of the city, and a new footbridge called Kaiser Bridge allowed me to make the same crossing as over the old High Bridge. Ever the mathematician, my immediate thought was whether I could make a journey around today's bridges in the spirit of the eighteenth-century game.

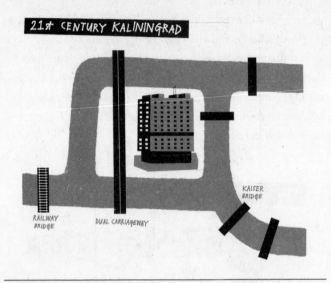

Figure 3.16   The bridges of Kaliningrad in the twenty-first century.

Euler's mathematical analysis told me that if there were exactly two places with an odd number of bridges emerging from them, there would be an Eulerian path: you start at one of the odd-numbered points and end at the other. By checking the plan of today's bridges of Kaliningrad, I found that such a journey is in fact possible.

The story of the bridges of Königsberg is important because it gave mathematicians a new way of looking at geometry and space. Rather than focusing on distances and angles, this new perspective concentrated on how shapes are connected together. This was the birth of topology (which we explored in chapter 2), one of the most influential branches of mathematics studied in the past hundred years. The problem of the bridges of Königsberg gave rise to the mathematics that currently runs modern Internet search engines, like Google's, which seek to maximize the way networks can be navigated.

WHAT'S THE MILLION-DOLLAR PROBLEM?

There are many different versions of this chapter's million-dollar problem. The classic one is called the traveling-salesman problem. An example of the problem is this challenge: you are a salesman and need to visit 11 clients, each located in a different town, and the towns are connected by roads, as shown in the following map—but you only have enough fuel for a journey of 238 miles.

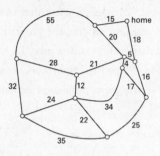

Figure 3.17   An example of the traveling-salesman problem. Can you find a route of 238 miles or less that visits all the points on this map and returns to the starting point?

The distance between towns is given by the number on the road joining them. Can you find a journey that lets you visit all 11 clients and then return home without running out of fuel? (The solution is at the end of the chapter.) In this version of the problem, the million dollars is on offer for a general algorithm or computer program that will produce the shortest path for any map you feed into the program that would be significantly quicker than getting the computer to carry out an exhaustive search. The number of possible journeys grows exponentially as you increase the number of cities, so an exhaustive search soon becomes practically impossible. Or can you prove that no such program is possible?

The general feeling among mathematicians is that problems of this sort have a built-in complexity, which means that there won't be any clever way to find the solution. I like to call these problems "needle in a haystack" problems because essentially, there are a vast number of possible solutions, and you're trying to find one in particular. The technical name for them is NP-complete problems.

One of the key characteristics of these puzzles is that once you've found the needle, it's easy to check that it does the job. For example, you know immediately once you've found a route around the map that is less than 238 miles long. A P-problem is one for which there is an efficient program for finding the solution. The million-dollar question can be put this way: are NP-complete problems in fact P-problems? Mathematicians refer to this as NP vs. P.

There is another very curious property that connects all these NP-complete problems. If you find an efficient program that works for one problem, it means that there will be such a program for all the other problems. To give an example of how this might work, here are two other "needle in a haystack," or NP-complete, problems that look very different.

The Diplomatic-Party Problem

You want to invite your friends to a party, but some of them can't stand one other—and you don't want two enemies in the same room. So you decide to have three parties and invite different people to each one. Can you send out invitations in such a way that two enemies won't turn up at the same party?

The Three-Color Map Problem

In chapter 2, we saw how you can always color a map with, at most, four colors. But is there an efficient way to tell whether you could get away with just three colors for any map?

How would a solution to the three-color map problem help you with the diplomatic-party problem? Let's say you've written down the names of your friends and drawn a line between pairs of people who *hate* each other:

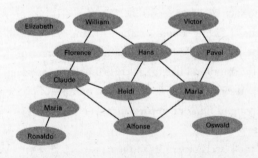

Figure 3.18   A line joins two people if they can't be invited to the same party.

To invite each friend to one of three parties, you could start by coloring the circles using three colors, each color corresponding to a different party. Deciding which friend to invite to which party is then the same as finding a way to color in this picture so that no connected friends have the same color. But look what happens when you replace the names of the friends with something different:

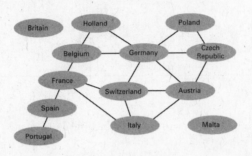

Figure 3.19   A line joins two countries if they share a common border.

Friends of yours who don't like one another have become European countries, and the links are now shared borders between countries. The problem of choosing which of the three parties to invite friends to has become the problem of choosing which of three colors to use for countries on a map of Europe.

The diplomatic-party question and the three-color map problem are different versions of the same question, so if you find an efficient way to solve one NP-complete problem, you end up solving them all! Here is a selection of different problems on which you can try your hand at winning the million.

Minesweeper

This is a single-player computer game that comes packaged with every copy of the Microsoft operating system. The aim of the game is to clear a grid of mines. If you click a square in the grid and it isn't a mine, you are shown how many of the surrounding squares contain mines. If you click on a mine . . . you've lost. But the million-dollar minesweeper challenge asks you to do something slightly different. The following picture can't be from a real game, because no arrangement of mines would give these numbers. The 1 implies that only one of the uncovered squares contains a mine, while the 2 implies that they both contain mines:

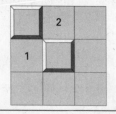

Figure 3.20

But what about the next picture—can it be from a real game?

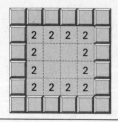

Figure 3.21

Is there a way to place mines so that the numbers are consistent? Or is there no way that this can be from a real game because there is no arrangement of mines that could give rise to the numbers showing? Your job is to come up with an efficient program that will work this out, whatever picture you are given.

Sudoku

Finding an efficient program to solve arbitrarily large sudoku puzzles is an NP-complete problem. Sometimes, with the really killer sudokus, you need to make some guesses and then follow through the logical implications of those guesses; there doesn't appear to be a clever way to make these guesses other than trying one after another until one set of guesses leads you to a consistent answer.

Packing Problem

You run a moving company. All your packing crates are of the same height and width, exactly the same as the internal dimensions of your truck (well, just a little smaller so that they just squeeze in). But the crates are of different lengths. Your truck is 150 feet long, and the crates available for packing have the following lengths: 16, 27, 37, 42, 52, 59, 65, and 95 feet.

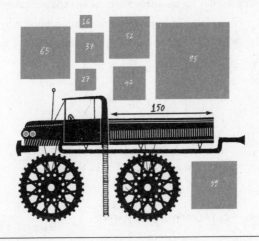

Figure 3.22   Packing crates is a mathematically complex problem.

Can you find a combination of crates that will fill the truck as efficiently as possible? You have to find an algorithm that decides—given any number N and a set of smaller numbers $n(1)$, $n(2)$, . . . , $n(r)$—whether there is a choice of the smaller numbers that add up to the big number.

These sorts of problems are not simply games: they crop up in business and industry when companies are faced with finding the most efficient solution to a practical problem. Wasted space or excess fuel costs companies money, and managers often need to solve one of these NP-problems. There are even some codes used in the telecommunications industry that depend on discovering the needle in the haystack to crack the code. So it's not just mathematicians or obsessive game players who are interested in cracking this million-dollar problem.

Whether it's mathematically analyzing routes for a travelling salesman or arranging parties, coloring maps, or sweeping mines, this chapter's million-dollar problem comes in so many different guises that there should be one version that appeals. But be warned: this problem may look like fun and games, but it is one of the hardest of all the million-dollar problems. Mathematicians believe that there is some essential complexity about all these problems that means there won't be an efficient program to solve them. The problem is that it's always more difficult to show why something *doesn't* exist than to show that it does. But at least you'll have fun trying to win this chapter's million.

SOLUTIONS

Number Mysteries Lottery

The winning numbers were 2, 3, 5, 7, 17, 42.

The Traveling-Salesman Problem

Here's a route that covers 238 miles:

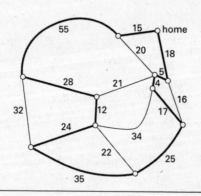

Figure 3.23

$$15 + 55 + 28 + 12 + 24 + 35 + 25 + 17 + 4 + 5 + 18 = 238$$

Four

THE CASE OF THE UNCRACKABLE CODE

Ever since people first learned to communicate, they've been finding ever more fiendish ways to hide messages from their enemies. Maybe you used a private code to write your diary so that your brother or sister couldn't read it, just as Leonardo da Vinci did. But codes aren't only for keeping things secret: they also make sure that information is communicated without errors. And we can use mathematics to create ingenious ways to guarantee that the message that is received is the same as the message that was sent—which is vitally important in this age of electronic transactions.

A code is simply a systematic way of arranging a set of symbols to convey a particular meaning. As soon as you start looking for codes, you discover that they are all around us: there are bar codes on everything we buy; codes enable us to store music in our MP3 players; codes let us browse the Internet. Even this book is written in code—English is simply a code formed with the 26 letters of the alphabet, and our "acceptable code words" are stored in the *Oxford English Dictionary.* Even our body contains codes—DNA is a code for reproducing a living creature made of four organic chemical substances called bases: adenine, guanine, cytosine, and thymine, or A, G, C, and T for short.

In this chapter, I'll show you how math has been used to create and break some of the cleverest codes around; how it lets us transmit information safely, efficiently, and secretly; and how it lets us do everything from photographing the planets from spacecraft to shopping on eBay. And at the end of the chapter, I'll explain how cracking one of the million-dollar math problems could help you crack codes as well.

HOW TO USE AN EGG TO SEND A SECRET MESSAGE

In Italy in the sixteenth century, Giovanni Porta discovered that you could write a hidden message on a hard-boiled egg with an ink made by dissolving an ounce of alum in a pint of vinegar. The ink penetrates the shell and marks the hardened white inside, but while the writing stays on the egg white, it disappears from the shell—perfect for sending secret messages: to crack the code, you crack the egg! And this is just one of the many crazy ways people have come up with to hide secret messages.

In 499 BC, a tyrant by the name of Histiaeus wanted to send a secret message to his nephew Aristagoras, encouraging him to rebel against the Persian king. Histiaeus was stationed in Susa, in modern-day Iran, and his nephew was back home in Miletus, now a part of Turkey. How could he get a message to his nephew without the Persian authorities intercepting it? He came up with a cunning plan. He shaved the head of his faithful servant and tattooed the message onto his bald head. Once the hair had grown back, Histiaeus sent the servant off to find his nephew. When the servant reached Miletus, the nephew shaved the servant's head, read the message, and began a rebellion against the ruling Persian king.

While the nephew only had to shave the messenger's head to retrieve the message, we have to pity the recipients of secret messages sent by an ancient Chinese method. The message would be written on a piece of silk, which was then rolled up very tightly. The silk ball was covered in wax and given to the messenger to swallow. Recovering the message once it had reappeared was not a particularly pleasant affair.

One of the most sophisticated ways of hiding a message was developed by the Spartans in 500 BC. They used a special wooden cylinder called a scytale around which they would wrap a thin strip of paper in a spiral. The secret message would then be written on the paper, lengthways down the scytale, but when the paper was unwrapped, the message looked like gobbledygook. It was only by winding the strip of paper around another scytale of exactly the same dimensions that all the letters lined up again correctly.

These methods for sending secret messages are examples of steganography—the art of concealing—rather than coding. But however devious they were, once the message was discovered, the secret was out. So people started thinking about how to conceal the meaning of the message even if it was uncovered.

HOW TO CRACK THE KAMA SUTRA CODES BY COUNTING

B OBDFSOBDLNLBC, ILXS B QBLCDSV MV B QMSD, LE B OBXSV MH QBDDSVCE.

LH FLE QBDDSVCE BVS OMVS QSVOBCSCD DFBC DFSLVE, LD LE ASNBGES DFSJ BVS OBTS ZLDF LTSBE. DFS OBDFSOBDLNLBC'E QBDDSVCE, ILXS DFS QBLCDSV'E MV DFS QMSD'E OGED AS ASBGDLHGI; DFS LTSBE ILXS DFS NMIMGVE MV DFS ZMVTE, OGED HLD DMUSDFSV LC B FBVOMCLMGE ZBJ. ASBGDJ LE DFS HLVED DSED: DFSVS LE CM QSVOBCSCD QIBNS LC DFS ZMVIT HMV GUIJ OBDFSOBDLNE.

This looks like gibberish, but it's a message written using one of the most popular codes ever created. Called the substitution cipher, it works by switching every letter of the alphabet with another one: so *a* might become *P, t* might become *C,* and so on. (I've used lowercase letters for those letters in the unencrypted message—the plaintext, as it's known in the trade—and capitals for the letters in the ciphertext.) As long as the sender and receiver have agreed to the substitution key in advance, the receiver will be able to decipher the messages, but to everyone else, they'll be a meaningless string of nonsense.

The simplest version of these codes is called a Caesar shift—after Julius Caesar, who used this type of code to communicate with his generals during the Gallic Wars. Caesar shifts work by shifting each letter the same number of places along. So with a shift of three, *a* becomes *D, b* becomes *E,* and so on. From the Number Mysteries website, you can download and cut out your own cipher wheel for creating these simple Caesar shifts.

Shifting each letter by the same number of places gives you only 25 possible ciphers, so once you've spotted that this is how the message has been encoded, it's quite easy to unscramble. There's a better way to encode a message: instead of simply shifting all the letters along together, we can mix things up and allow any letter to be substituted for any other. This method of encrypting messages was actually suggested some centuries before Julius Caesar, and surprisingly, it wasn't in a military handbook but in the Kama Sutra. Although this ancient Sanskrit text is usually associated with a description of physical pleasures, it covers a host of other arts that the author believed a woman should be versed in, from conjuring and chess to bookbinding and even carpentry. The forty-fifth chapter is dedicated to the art of secret writing, and here, the substitution cipher is explained as a perfect way of concealing messages detailing liaisons between lovers.

While there are only 25 Caesar shifts, if we allow ourselves to substitute any letter for any other, then the number of possible ciphers is somewhat greater. We have 26 choices for what to change the letter *a* into, and for each of those possibilities, there are 25 choices of where to send the letter *b* (one letter has already been used to encrypt the letter *a*). So there are already 26 × 25 different ways to scramble the letters *a* and *b*. If we keep going, choosing different letters for the rest of the alphabet, we find that there are

$$26 \times 25 \times 24 \times 23 \times 22 \times 21 \times 20 \times 19 \times 18 \times 17 \times 16 \times 15 \times 14 \times 13 \times$$
$$12 \times 11 \times 10 \times 9 \times 8 \times 7 \times 6 \times 5 \times 4 \times 3 \times 2 \times 1$$

different Kama Sutra codes. As we saw on page 39, we can write this as 26! We should also remember to take 1 off this number because one choice will have been *A* for *a*, *B* for *b*, and so on through *Z* for *z*, which isn't much of a code. When we multiply out 26! and take away 1, we arrive at the grand total of

403,291,461,126,605,635,583,999,999

different ciphers—over four hundred million billion billion possibilities.

The passage at the beginning of this section is encrypted with one of those codes. To give you an idea of how many possible permutations there are, if I wrote out the passage using all the different possible codes, then the piece of paper would stretch from here to well beyond the outskirts of our Milky Way Galaxy. A computer checking one code a second since the big bang went off over thirteen billion years ago would still be just a fraction of the way through checking each of the codes—and a very small fraction at that.

So the code looks virtually uncrackable. How on earth (or beyond) can you find out which of this vast number of possible codes I've used to encrypt my message? Amazingly, the way to do it is by a very simple bit of mathematics: counting.

	a	b	c	d	e	f	g	h	i	j	k	l	m
%	8	2	3	4	13	2	2	6	7	0	1	4	2

	n	o	p	q	r	s	t	u	v	w	x	y	z
%	7	8	2	0	6	6	9	3	1	2	0	2	0

Table 4.1

It was the Arabs in the time of the Abbasid dynasty who first developed cryptanalysis, as the science of code breaking is called. The ninth-century polymath Ya'qub al-Kindi recognized that in a piece of written text, some letters crop up again and again, while others are used rarely, as shown in the table. This is something that Scrabble players are well aware of: the letter *e* is worth only one point, as it is the most common letter in the English alphabet, while a *z* is worth ten points. In written texts, every letter has its own distinct "personality"—how often it appears, and in what combinations with others—but the key to al-Kindi's analysis is to realize that a letter's personality doesn't change when it is represented by another symbol.

So let's make a start at cracking the code used to encrypt the text at the beginning of this section. Here's a breakdown of the frequency of each of the letters used in the encrypted text:

	A	B	C	D	E	F	G	H	I	J	K	L	M
%	1	10	5	12	7	6	3	2	2	1	0	8	5

	N	O	P	Q	R	S	T	U	V	W	X	Y	Z
%	2	4	0	3	0	13	1	1	7	0	1	0	1

Table 4.2

From the graph, we can see that the letter *S* occurs with a frequency of 13 percent, more than any other letter in the ciphertext, so there's a good chance that this letter is the one used to encode the letter *e*. (Of course, you have to hope that I haven't chosen a passage from Georges Perec's novel *A Void*, which was written without using the letter *e* anywhere in the text.) The next most common letter in the ciphertext is *D*, with a frequency of 12 percent. The second most common letter in the English language is *t*, so this would be a good guess for the choice of *D*. Third in order of frequency in the encrypted message is *B*, which is used 10 percent of the time, so there is a strong possibility that it stands for the third most common letter in the English language: *a*.

Let's substitute these letters into the text and see what we get:

a OatFeOatLNLaC, ILXe a QaLCteV MV a QMet, LE a OaXeV MH QatteVCE.

LH FLE QatteVCE aVe OMVe QeVOaCeCt tFaC tFeLVE, Lt LE AeNaGEe tFeJ aVe OaTe ZLtF LTeaE. tFe OatFeOatLNLaC'E QatteVCE, ILXe tFe QaLCteV'E MV tFe QMet'E OGEt Ae

AeaGtLHGI; tFe LTeaE ILXe tFe NMIMGVE MV tFe ZMVTE, OGEt
HLt tMUetFeV LC a FaVOMCLMGE ZaJ. AeaGtJ LE tFe HLVEt
teEt: tFeVe LE CM QeVOaCeCt QIaNe LC tFe ZMVIT HMV GUIJ
OatFeOatLNE.

You might say that this still looks like gobbledygook, but the fact that
you see the letter *a* on its own several times is telling us that we've probably
decoded this letter correctly. (*B* might turn out to stand for *i*, of course, in
which case we'd have to go back and try again.) And we can see that the
word *tFe* occurs quite often, so it's a good bet that this is the word *the*. Indeed, the letter *F* accounts for 6 percent of the ciphertext, and the letter *h*
occurs 6 percent of the time in English.

We can see the word *Lt*, in which only the second letter has been decoded. There are only two two-letter words that end in *t*: *at* and *it*. We've
already decoded *a,* so there's a good chance that *L* should be decoded as *i*, and
our frequency chart bears this out. *L* appears in the ciphertext with 8 percent
frequency, and *i* appears in the English language with 7 percent frequency—a
close match. This isn't an exact science: the longer the text, the more in tune
the frequencies will be, but we have to be flexible when we use this technique.

Let's put in our two new decodings:

a OatheOatiNiaC, IiXe a QaiCteV MV a QMet, iE a OaXeV MH
QatteVCE.

iH hiE QatteVCE aVe OMVe QeVOaCeCt thaC theiVE, it iE
AeNaGEe theJ aVe OaTe Zith iTeaE. the OatheOatiNiaC'E
QatteVCE, IiXe the QaiCteV'E MV the QMet'E OGEt Ae
AeaGtiHGI; the iTeaE IiXe the NMIMGVE MV the ZMVTE, OGEt
Hit tMUetheV iC a haVOMCiMGE ZaJ. AeaGtJ iE the HiVEt
teEt: theVe iE CM QeVOaCeCt QIaNe iC the ZMVIT HMV GUIJ
OatheOatiNE.

Gradually, the message is beginning to emerge. I'll leave to you the task
of unraveling the rest; the decoded text is at the end of the chapter if you

want to check whether you're right. I'll give you a hint: it's a couple of my favorite passages from *A Mathematician's Apology* by the Cambridge mathematician G. H. Hardy. I read this book when I was in school, and it was one of the things that made me decide to become a mathematician.

This simple mathematical trick of counting letters means that any message disguised with a substitution cipher cannot be made secret—as Mary Queen of Scots found to her cost. Using a code that substituted strange symbols for the letters, she wrote messages to her fellow conspirator, Anthony Babington, about their plans to assassinate Queen Elizabeth I:

Figure 4.1   The Babington code.

At first glance, the messages Mary sent looked impenetrable, but Elizabeth had at her court one of the master code breakers of Europe: Thomas Phelippes. He was not a handsome man, as the following description of him makes clear: "of low stature, slender every way, dark yellow haired on the head and clear yellow beard, even in the face with smallpox, of short sight." Many people believed that Phelippes had to be in league with the devil to read such hieroglyphs, but his trick was to apply the same principle of frequency analysis. He cracked the code, and Mary was arrested and put on trial. The decoded letters were the evidence that ultimately led to her execution for conspiracy.

If you're faced with cracking a Kama Sutra code, the following web page is helpful in analyzing the frequency of different letters in the encrypted text: www .simonsingh.net/The_Black_Chamber/ frequencypuzzle.htm.

You can access it with your smartphone by scanning this code.

HOW DID MATHEMATICIANS HELP TO WIN THE SECOND WORLD WAR?

Once it was known that substitution codes had this inherent weakness, cryptographers started to devise more ingenious ways to thwart attacks based on counting letters. One idea was to vary the substitution cipher. Instead of using just one substitution cipher to encode the whole text, you could alternate between two different ciphers. Then, if you were encoding the word *beef,* say, the letter *e* would be encoded differently each time, because the first one would be encoded using one cipher, and the second one by another cipher. So *beef* might get encoded as *PORK.* The more secure you want your message to be, the more ciphers you can cycle through.

Of course, there is a balance to be struck in cryptography between making things very secure and having a cipher that is usable. The most secure sort of cipher, called a one-time pad, uses a different substitution cipher for every single letter of the text. It is almost uncrackable because there is absolutely nothing to help you to come to grips with the encrypted text. It is also very unwieldy because you need to use a different substitution cipher for each letter in the message.

The sixteenth-century French diplomat Blaise de Vigenére believed that to hinder any frequency analysis, it was enough to cycle through just a few substitution ciphers. Although the Vigenére code, as it became known, was a much stronger form of encryption, it was not unbreakable, and the British mathematician Charles Babbage eventually found a way to crack it.

Babbage is regarded as the grandfather of the computer age for his belief that machines can be used to automate calculations; a reconstruction of his "difference engine" calculating machine can be seen in London's Science Museum. It was his systematic approach to problems that, in 1854, gave him the idea for cracking the Vigenére code.

Babbage's method depends on one of the great skills of the mathematician—pattern recognition. The first thing you need to spot is how many different substitution ciphers are being cycled through. Because the word *the* will usually appear often in any plaintext message, spotting repetitions of the same three-letter sequence can be the key to unraveling how many ciphers are being used. So, for example, you might spot *AWR* appearing frequently, and that there is always a gap of a multiple of four letters between occurrences of *AWR*. That would be a good indication that four ciphers are being used.

Once you have this information, you can split the ciphertext into four groups. The first group consists of the first letter, fifth letter, ninth letter, and so on. The second group consists of the second letter, sixth letter, tenth letter, and so on. The same substitution cipher will have been used to encode the letters in any one of these four groups, so you can now use frequency analysis on each group in turn, and the code is cracked.

As soon as the Vigenére code was broken, the search was on for a new way to encode messages securely. When the Enigma machine was developed in Germany in the 1920s, many people believed that the ultimate uncrackable code had been created.

The Enigma machine works on the principle of changing the substitution cipher each time a letter is encoded. If I want to encode the sequence *aaaaaa* (to indicate perhaps that I'm in pain), then each *a* will get encoded in a different way. The beauty of the Enigma machine was that it mechanized the change from one substitution to the next very efficiently. The message is typed on a keyboard. There is a second bank of letters—the "light board" above the keyboard—and when a key is pressed on the keyboard, one of the letters above lights up to indicate the encoded letter. But the keyboard isn't wired directly to the light board: the connections are via three disks that contain a maze of wiring and can be rotated.

One way to understand how an Enigma machine works is to imagine a large cylinder consisting of three rotating drums. At the top of the cylinder are 26 holes around the rim, labeled with the letters of the alphabet. To encode a letter, you drop a ball into the hole corresponding to that letter. The ball drops into the first drum, which has 26 holes around the rim at the top and 26 holes around the rim at the bottom. Tubes connect the upper and lower holes—but they don't simply connect the holes at the top to the holes directly below them. Instead, the tubes twist and turn, so that a ball entering the drum at the top will pop out of a hole at the bottom in a completely different location. The middle and lower drums are similar, but their connecting tubes link the holes at the top with the holes at the bottom in different ways. When the ball drops out of a hole at the bottom of the third drum, it enters the last piece of the contraption and emerges from one of 26 holes at the bottom of the cylinder, each of which is, again, labeled with the letters of the alphabet.

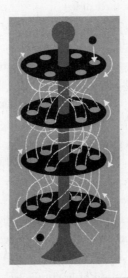

Figure 4.2   *The principle behind the Enigma machine: drop a ball through the tubes to encode a letter. The cylinders rotate after each encoding, so the letters are scrambled differently each time.*

Now, if our contraption just stayed as it was, it would be nothing more than a complicated way of reproducing a substitution cipher. But here's the genius of the Enigma machine: every time a ball drops through our cylinder, the first drum rotates by ½6th of a turn. So when the next ball is dropped through, the first drum will send it on a completely different route. For example, while the letter *a* might first be encoded by the letter *C*, once the first drum moves on one step, a ball dropped in the letter *a* hole will emerge from a different hole at the bottom. And so it was with the Enigma machine: after the first letter had been encoded, the first rotating disk clicked around by one position.

The rotating disks work a bit like an odometer: once the first disk has clicked around through all 26 positions, as it returns to its starting position, it moves the second disk on by ½6th of a turn. So in all, there are 26 × 26 × 26 different ways to scramble the letters. Not only that, but the Enigma operator could also alter the order of the disks, multiplying the number of possible substitution ciphers by a factor of 6 (corresponding to the 3! different ways of arranging the three disks).

Each operator had a codebook that described how, at the start of each day, the three disks should be arranged to encode messages. The recipient would decode the message with the same setting from the codebook. More complexities were introduced into the way the Enigma machine was constructed, and there were ultimately over 158 million million million different ways to set up the machine.

In 1931, the French government discovered the plans for the German machine, and they were horrified. There seemed to be no possible way to work out from an intercepted message how the disks were set up for each day's encoding—which was crucial if a message was to be decoded. But they had a pact with the Poles to exchange any intelligence that was gathered, and the threat of German invasion had the effect of concentrating Polish minds.

Mathematicians in Poland realized that each setting of the disks had its own particular features, and that patterns could be exploited to help work backward and crack encrypted messages. If the operator typed an *a*, for example, it would be encoded according to how the disks were set—let's say as *D.* The first disk then clicks on one step. If, when another *a* is typed, it's encoded as *Z,* then in some sense, *D* is related to *Z* by the way the disks have been set.

We could investigate this using our contraption. By resetting the drums and dropping balls twice through each letter in turn, we would build up a complete set of relationships that might look like this:

Letter to code	a	b	c	d	e	f	g	h	i	j	k	l	m
1st ball	D	T	E	R	F	A	Q	Y	S	I	P	B	N
2nd ball	Z	S	B	Q	X	G	L	V	K	A	J	D	Y

Letter to code	n	o	p	q	r	s	t	u	v	w	x	y	z
1st ball	C	G	Z	J	H	M	U	X	K	O	W	L	V
2nd ball	H	C	W	E	O	I	M	T	P	F	N	R	U

Table 4.3

Each letter appears once and once only in each row because each row corresponds to a single substitution cipher.

How did the Poles use these relationships? On any given day, all the German Enigma operators would be using the same setting for the wheels, which they would find in their codebook. They would then choose their own setting, which they would then send using the original setting from the codebook. To be on the safe side, they were encouraged to repeat their choice by typing it twice. Far from being safe, this turned out to be a fatal mistake. It gave the Poles a hint about how the wheels connected the letters—a clue as to how the Enigma machine was set up for that day.

A group of mathematicians based in a country house at Bletchley Park, halfway between Oxford and Cambridge, studied the patterns that the mathematicians in Poland had spotted and found a way to automate the search for the settings using a machine they built that was known as a bombe. It's been said that those mathematicians shortened the Second World War by two years, saving countless lives. And the machines they built ultimately gave birth to the computers we all rely on today.

 *For an online simulation of the Enigma machine, check out www.bletchleypark. org.uk/content/enigmasim.rhtm, which you can access directly by scanning this code with your smartphone.*

From the Number Mysteries website, you can download a PDF of instructions for making your own Enigma machine.

GETTING THE MESSAGE ACROSS

Whether your message is encoded or not, you still need to find a way to transmit it from one location to another. Many ancient cultures, from the ancient Chinese to the Native Americans, used smoke signals as a way of communicating over long distances. It's said that the fires that were lit on the towers of the Great Wall of China could communicate a message for three hundred miles along the wall in a matter of hours.

Visual codes based on flags date back to 1684, when Robert Hooke, one of the most famous scientists of the seventeenth century, suggested the idea to the Royal Society in London. The invention of the telescope had made it possible to communicate visual signals over large distances, but Hooke was spurred on by something that has led to many new technological advances: war. The previous year, the city of Vienna had almost been captured by the Turkish army without the rest of Europe knowing. Suddenly, it was a matter of urgency to come up with a way of sending messages quickly over large distances.

Hooke proposed setting up a system of towers right across Europe. If one tower sent a message, it would be repeated by all the other towers within visual range—a two-dimensional version of the way messages were sent down the length of the Great Wall of China. The method of transmitting messages wasn't very sophisticated: large characters would have to be hoisted aloft on ropes. Hooke's proposal, however, was never implemented, and it was another hundred years before a similar idea was put into practice.

In 1791, the brothers Claude and Ignace Chappe built a system of towers to speed up the French Revolutionary government's communications (though one tower was attacked when mobs thought it was actually the Royalists who were communicating with each other). The idea came from a system that the brothers had used to send messages between dormitories at the strict school where they had boarded as children. They experimented with lots of different ways of sending messages visually, and in the end, they settled on wooden rods set at different angles, which the human eye could easily distinguish.

Figure 4.3   The Chappe brothers' code was transmitted via hinged wooden arms.

The brothers developed a code based on a movable system of hinged wooden arms to denote different letters or common words. The main cross arm could be set at four different angles, while two smaller arms attached to the end of the cross arm had seven different settings, making it possible to communicate a total of $7 \times 7 \times 4 = 196$ different symbols. Although part of the code was used for public communication, 92 of the symbols, combined in pairs, were used by the brothers for a secret code, representing 92 × 92 = 8,464 different words or phrases.

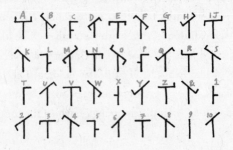

Figure 4.4   Letters and numbers as transmitted by the Chappe brothers' communication system.

In their first test, on March 2, 1791, the Chappe brothers successfully sent the message "If you succeed, you will soon bask in glory" across a distance of ten miles. The government was sufficiently impressed with the brothers' proposal that in four years, a system of towers and flags was constructed that stretched right across France. In 1794, one line of towers, covering a distance of 143 miles, successfully communicated the news that the French had captured the town of Condé-sur-l'Escaut from the Austrians less than an hour after it had happened. Unfortunately, success did not lead to the glory that very first message had predicted. Claude Chappe got so depressed when he was accused of plagiarizing existing telegraph designs that he ended up drowning himself in a well.

It was not long before flags began to replace wooden arms on the tops of towers, and flags were adopted by sailors for communicating at sea, since all they had to do was wave them from a position visible to other ships. Perhaps the most famous coded message to be sent between ships using flags was this one, sent at 11:45 on October 21, 1805:

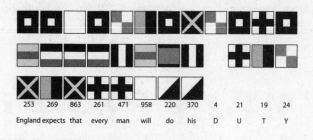

Figure 4.5   Admiral Nelson's famous message.

This was the message that Horatio Nelson hoisted on his flagship HMS *Victory* just before the British navy engaged in the decisive clash that won the Battle of Trafalgar. The navy was using a secret code developed by another admiral, Sir Home Popham. Codebooks were distributed to each of the ships in the navy and were lined with lead so that if the ship was taken, the codebook could be thrown overboard to stop the enemy from capturing the British secret cipher.

The code worked by using combinations of ten different flags in which each flag represented a different numeral from 0 to 9. The flags would be run up the mast of a ship three at a time, indicating a number from 000 to 999. The recipient of the message would then look in the codebook to see what word was encoded by that number. *England* was encoded by the number 253, and the word *man* by 471. Some words, such as *duty*, were not in the codebook and would need to be spelled out by flags reserved for individual letters. Originally, Nelson had wanted to send the message "England *confides* that every man will do his duty," in the sense that England was confident, but the signal officer, Lieutenant John Pasco, couldn't find the word *confides* in the codebook. Rather than spelling it out, he politely suggested to Nelson that maybe *expects,* which was in the book, was a better word.

The use of flags was overtaken by the development of telecommunications, but the modern system, using a flag held in each hand, is still learned by sailors today and is known as semaphore. There are eight different positions for each arm, making 8 × 8 = 64 possible different symbols.

Figure 4.6   Semaphore.

Nujv!

NUJV!

Figure 4.7

On the front cover of their album Help!, *the Beatles are apparently using semaphore to announce the title. But though* they are making semaphore signs, when you decode the message, it doesn't read HELP but NUJV. Robert Freeman, who had the idea of using semaphore on the cover, explained that "when we came to do the shot, the arrangement of the arms with those letters didn't look good. So we decided to improvise and ended up with the best graphic positioning of the arms."

They should have been doing this:

Figure 4.8

The Beatles aren't the only band to have used codes incorrectly on an album cover, as we shall see.

 To see how a message translates into semaphore, check out http://inter.scoutnet .org/semaphore or scan this code with your smartphone.

Figure 4.9   Did you know that the peace symbol used by the Campaign for Nuclear Disarmament is actually semaphore? It represents the letters n *and* d *combined into one symbol.*

WHAT IS THE CODED MESSAGE IN BEETHOVEN'S FIFTH SYMPHONY?

Beethoven's Fifth Symphony begins with one of the most famous openings in the history of music—three short notes followed by a long note. But why, during the Second World War, did the BBC start every radio broadcast of the news with Beethoven's famous motif? The answer is that it contains a coded message. This new code exploited technology that could send signals through wires in a series of electromagnetic pulses.

One of the first to experiment with this form of communication was Carl Friedrich Gauss, whose work on prime numbers we looked at in chapter 1. As well as mathematics, he was also interested in physics, including the emerging field of electromagnetism. He and the physicist Wilhelm Weber rigged up a wire a kilometer in length running from Weber's laboratory in Göttingen to the observatory where Gauss lived, and used it to send messages to each other.

To do this, they needed to develop a code. At each end of the wire, they set up a needle attached to a magnet that the wire was wrapped around. By changing the direction of the current, the magnet could be made to turn left or right. Gauss and Weber devised a code that turned letters into combinations of left and right turns:

r = a	rrr = c, k	lrl = m	lrrr = w	llrr = 4
l = e	rrl = d	rll = n	rrll = z	lllr = 5
rr = i	rlr = f, v	rrrr = p	rlrl = 0	llrl = 6
rl = o	lrr = g	rrrl = r	rllr = 1	lrll = 7
lr = u	lll = h	rrlr = s	lrrl = 2	rlll = 8
ll = b	llr = l	rlrr = t	lrlr = 3	llll = 9

Table 4.4

Weber was so excited by the potential of their discovery that he prophetically declared, "When the globe is covered with a net of railroads and telegraph wires, this net will render services comparable to those of the nervous system in the human body, partly as a means of transport, partly as a means for the propagation of ideas and sensations with the speed of light."

Many different codes were suggested to fulfill the potential of electromagnetism to communicate messages, but the code developed by the American Samuel Morse in 1838 was so successful that it put all the others out of business. It was similar to Gauss and Weber's scheme, converting each letter into a combination of long and short bursts of electricity: dashes and dots.

The logic on which Morse based his code is something like the frequency analysis used by code breakers to crack the substitution cipher. The most common letters in the English alphabet are *e* and *t*, so it makes sense to use the shortest possible sequence to encode them. So *e* is represented by a dot (a short burst of electricity), and *t* is represented by a dash (a long burst). The less common letters require longer sequences, so *z*, for example, is dash-dash-dot-dot.

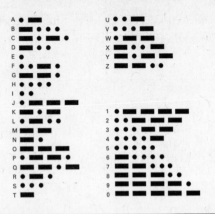

Figure 4.10   Morse code.

With the aid of Morse code, we can now crack the message hidden in Beethoven's Fifth. If we interpret the dramatic opening of the piece as Morse code, then dot-dot-dot-dash is Morse for the letter *v*, which the BBC used to symbolize victory.

Although Beethoven certainly didn't intend to hide messages in Morse in his music, given that he died before it was invented, other composers have deliberately used rhythm to add an extra layer of meaning to their work. The music for the famous detective series *Inspector Morse* appropriately enough begins with a rhythm that spells out the detective's name in Morse code:

Figure 4.11

In some of the episodes, the composer even threaded the name of the story's killer in Morse during the incidental music accompanying the program, though red herrings sometimes found their way into the score.

Although Morse code has been used extensively—not just by composers but by telegraph operators the world over—there is an inherent problem. If you receive a dot followed by a dash, how should you decode it? This is Morse for the letter *a*—but it is also Morse for the letter *e* followed by the letter *t*. As a result, mathematicians have found that a different sort of code, using 0s and 1s, is much more suitable for machines to understand.

WHAT IS THE NAME OF COLDPLAY'S THIRD ALBUM?

When fans rushed out to buy Coldplay's third album, released in 2005, there was a lot of excitement over the meaning of the graphics on the front cover, which depicted various colored blocks arranged in a grid. What was the significance of the picture? It turned out to be the title of the album written in one of the very first binary codes, proposed in 1870 by a French engineer, Émile Baudot. The colors were irrelevant: what mattered was that each block represents a 1, and gaps are to be read as 0s.

The seventeenth-century German mathematician Gottfried Leibniz was one of the first to realize the power of 0s and 1s as an effective way of coding information. He got the idea from the Chinese book *I Ching—Book of Changes*—which explores the dynamic balance of opposites. It contains a set of 64-line arrangements known as hexagrams, which are meant to represent different states or processes, and it was these that inspired Leibniz to create the mathematics of binary (which we met in the last chapter, when we looked at how to win at nim). The symbols consist of a stack of six horizontal lines in which each line in the symbol is either solid or broken. The *I Ching* explains how these symbols can be used in divination by tossing sticks or coins to determine the structure of the hexagram.

For example, if the fortune-teller cast the hexagram

Figure 4.12

it would indicate "conflict." But if the lines came out the other way around and you got

Figure 4.13

that would indicate "hidden intelligence."

Leibniz was more interested in the fact that Shao Yong, a Chinese scholar of the eleventh century, had pointed out that each symbol could be attached to a number. If you write 0 for a broken line and 1 for a solid line, then reading the first hexagram from top to bottom would give you 111010. In decimal numbers, each position corresponds to a power of 10, and the number in that position tells you how many powers of 10 to take. So 234 is 4 units, 3 tens, and 2 hundreds.

But Leibniz and Shao Yong weren't working in decimal, but binary, where each position is a power of 2. In binary, the number 111010 stands for no units, one lot of 2, no 4s, one lot of 8, one lot of 16, and one lot of 32. Add these up and the total is 2 + 8 + 16 + 32 = 58. The beauty of binary is that you need only two symbols, rather than the ten in decimal, to represent any number. Two lots of (decimal) 16 become one lot of the next power of 2, which is 32.

Leibniz saw that this way of representing numbers was very powerful if you wanted to mechanize calculations. There are very simple rules for adding binary numbers. At each position, 0 + 1 = 1, 1 + 0 = 1, and 0 + 0 = 0. The fourth possibility is 1 + 1 = 0, with the secondary effect that a 1 is carried over and added to the next position to the left. When we add 1000 to 111010, for example, we get a domino effect as the 1s cascade up the number:

1000 + 111010 = 10000 + 110010 = 100000 + 100010 = 1000000 + 000010 = 1000010

Leibniz designed beautiful mechanical calculators. One of them used ball bearings to represent 1s and an absence of balls to represent 0s, turning the process of addition into a fantastic mechanical pinball machine. Leibniz believed that "it is unworthy of excellent men to lose hours like slaves in the labour of calculation which would safely be relegated to anyone else if machines were used." I think most mathematicians would agree.

Figure 4.14   A reconstruction of Leibniz's binary calculator.

It wasn't only numbers that people started to represent as strings of 0s and 1s, but letters, too. Although humans found Morse code a very powerful tool for communicating, machines were less adept at picking up the subtle differences between dots and dashes spelling out letters and knowing when one letter had finished and the next had begun.

In 1874, Émile Baudot proposed coding each letter of the alphabet as a string of five 0s and 1s. By making every letter the same length, it was clear where the last letter finished and the next one began. Using five 0s and 1s allowed Baudot to represent a total of $2 \times 2 \times 2 \times 2 \times 2 = 32$ different characters. The letter x became the string 10111, for example, while y was 10101. This was a huge breakthrough, because messages could now be coded onto a paper tape on which holes were punched to represent 1s and blanks—the absence of a hole—represented 0s. A machine could read this tape and send the signal down a wire at high speed, and a teleprinter at the other end could automatically type out the message.

The Baudot code has since been superseded by a whole host of other codes that exploit the same idea of using 0s and 1s to represent everything from text to sound waves, and from JPEGs to movie files. Every time you log on to iTunes and download a Coldplay track, your computer receives a huge onslaught of 0s and 1s, which your MP3 player knows how to decode. Contained inside the numbers are the messages to tell your speaker or headphones how to vibrate so that you can hear Chris Martin's dulcet tones. And it may have been the fact that in our digital age, the music is simply a stream of 0s and 1s that inspired the cover of Coldplay's third album.

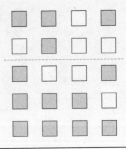

Figure 4.15   *The cover of Coldplay's third album used the Baudot code.*

Baudot's original code is the key to unlocking the secret message embedded in the cover's graphic. The pattern can be divided into four columns with five blocks in each column. The colored blocks should be interpreted as 1s, and the gaps as 0s. Because it is sometimes hard to tell which way up the tape should be, the machine punches a fine dividing line between the upper two blocks and the three below. That's why there's a line separating the gray blocks and the colored blocks on the cover's graphic.

The first column on the cover reads color-blank-color-color-color, which translates into 10111, the Baudot code for *x*. The last column becomes the Baudot code for *y*. The middle two columns are slightly more interesting. Five 0s and 1s gives you the possibility of 32 symbols, but quite often, we want more than this, since there are numbers, punctuation marks, and other symbols we might want to communicate. To cope with these demands, Baudot came up with a cunning way to expand the range. Just as a keyboard uses a shift key to get access to a whole range of other symbols using the same keys, Baudot used one of the string of five 0s and 1s as the equivalent of the shift key. So if you come across 11011, you know the next string will be from the extended set of characters.

And the second column on the cover is Baudot's shift key. To decode the blank-blank-blank-color-color in the third column, we need to consult the extended character set, shown in Figure 4.16. I'm sure most people will be expecting to find the symbol for &. But 00011 stands not for &, but for the numeral 9. So the real title of Coldplay's third album as encoded with the Baudot code is *X9Y*, not *X&Y*. Were the members of Coldplay playing a joke on us? Probably not. There is just one block difference between the Baudot code for 9 and &, so it was probably a mistake, which perfectly illustrates one of the problems with many of these codes: when you make a mistake, it's difficult to tell. It's in detecting errors such as this that the mathematics of codes truly comes into its own.

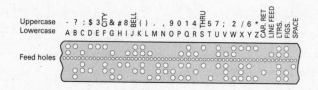

Figure 4.16   The Baudot code.

This website lets you create your own Coldplay album cover: www.ditonus.com/ coldcode. Or use your smartphone to scan this code.

WHICH OF THESE NUMBERS IS CODE FOR A BOOK: 0521447712 OR 0521095788?

I'm sure you've seen the ISBN—the International Standard Book Number—on the back of every book. In its ten digits, the ISBN uniquely identifies the book, as well as tells you its country of origin and its publisher. But that's not all this code does. The ISBN has a little magic worked into it.

Say I want to order a book and I know its ISBN. I type the number in, but I'm in a hurry and I make a mistake. You might think I'd end up with the wrong book, but that won't happen because ISBNs have an amazing property: they can detect errors in themselves. Let me show you how.

Here are some genuine ISBNs from some of my favorite books:

ISBN digit	0	5	2	1	4	2	7	0	6	1	Total
When multiplied	0	10	6	4	20	12	49	0	54	10	165

ISBN digit	1	8	6	2	3	0	7	3	6	9	Total
When multiplied	1	16	18	8	15	0	49	24	54	90	275

ISBN digit	0	4	8	6	2	5	6	6	4	2	Total
When multiplied	0	8	24	24	10	30	42	48	36	20	242

Table 4.5

Underneath each digit, I've multiplied the digit by its position in the code. So in the first ISBN, 0 gets multiplied by 1, 5 by 2, 2 by 3, and so on. Then I've added up all these new numbers and written the total at the end of the row. Notice anything about all these numbers that have been cooked out of the ISBN? Here are some more numbers you get by performing this calculation on other real ISBNs: 264, 99, 253.

Have you spotted the pattern? The calculation always gives a number that's divisible by 11. This is not an amazing coincidence, but an example of cunning mathematical design. It's only the first nine digits that contain the information about the book. The tenth digit is included just to make this total number extracted from the ISBN divisible by 11. You might have spotted that some books have an *X* instead of a number as their tenth digit. For example, another of my favorite books has ISBN 080501246X. The *X* actually stands for 10 (think roman numerals). In this case, you needed to add a multiple of 10 to the end of the ISBN to make the number you get from the ISBN divisible by 11.

If I get one of the digits wrong when I type out my ISBN, the calculation will give me a number that isn't divisible by 11, and the computer will know that I've made an error and ask me to type it in again. Even if I get two digits the wrong way around, which people often do when typing a number, it will also pick up this mistake, and rather than sending me the wrong book, it will ask me to send the correct ISBN. Pretty clever stuff. So now you can check which of the numbers in the title of this section is really the ISBN of a book and which one is the imposter.

With so many books continuing to be published, ISBNs were starting to run out. So it was decided that from January 1, 2007, ISBNs would have 13 digits. Twelve digits would once again identify the book, publisher, and country of publication, and the thirteenth would keep track of any errors that might creep in. But divisibility by 10 rather than 11 is the key to the 13-digit ISBNs now used by publishers. Look at the ISBN for this book, which you can find on the back. It has 13 digits. Add up the second, fourth, sixth, eighth, tenth, and twelfth digits and multiply the sum by 3. Now add on the other digits. The total will be divisible by 10. If you make a mistake in writing down the ISBN, the calculation will often give you a number not divisible by 10.

HOW TO USE CODES TO READ MINDS

You will need 36 coins to perform this trick. Give your unsuspecting friend 25 of the coins and ask him to arrange the coins in a 5 × 5 grid, with heads and tails showing at random. He might lay down the coins like this:

H	H	T	T	T
T	T	H	T	T
H	H	H	T	H
T	H	H	T	T
T	T	T	T	T

Table 4.6

Now you say, "In a minute, I'm going to ask you to turn over one of these coins—a head or a tail. Then I'll read your mind and tell you which of the coins you turned over. Now, I might be able to remember the order of the 25 coins, so let's make it even more difficult for me and make the square even bigger."

Next, you add more coins, apparently at random, to make an extra row and column and increase the grid to $6 \times 6 = 36$ . . . except that you aren't adding the extra coins randomly at all. What you do is count how many tails there are in each row and each column, starting with the first column. If there are an odd number of tails in this first column, then place the extra coin at the bottom of this column with a tail showing. If there are an even number of tails (0 counts as an even number for this purpose), place the extra coin at the end of this column with a head showing.

Do the same for each column, and then add a coin at the end of each row using the same criterion. There will now be a space at the bottom right to fill to complete the square. Make it a head or a tail according to whether the column above it has an even or odd number of tails. Interestingly, this will also record the parity (i.e., whether even or odd) of the number of tails in the bottom row, too. Can you prove that this is always true? The trick lies in spotting that this number is telling you whether there are an odd or even number of tails in the 5×5 grid.

Anyway, the grid now looks like this:

H	H	T	T	T	T
T	T	H	T	T	H
H	H	H	T	H	T
T	H	H	T	T	T
T	T	T	T	T	T
T	H	H	T	H	H

Table 4.7

Now you are ready to do the trick. Turn your back and ask your friend to turn over one coin so that it changes from a head to a tail or vice versa. When that's done, turn around again. Concentrate on the grid and announce that you're now going to read your friend's mind and identify the coin that was turned over.

Of course, you aren't reading your friend's mind at all. You go back to the original 5 × 5 block of numbers and count the heads and tails in each column and row. You note whether there are an odd or even number of tails and check it against the heads and tails you added, because they indicate the parity of tails in each column. Now that your friend has turned over one of the coins in the 5 × 5 grid, there will be one row and one column in which the coins you added give a false reading. Look at where that column and row intersect and you will find the coin that was turned over.

You should now be able to identify which coin has been changed in the grid:

H	H	T	T	T	T
H	T	H	T	T	H
H	H	H	T	H	T
T	H	H	T	T	T
T	T	T	T	T	T
T	H	H	T	H	H

Table 4.8

The first column in the 5 × 5 grid has an even number of tails, but the coin you added at the bottom of this column was a tail, indicating that there were originally an odd number of tails. So the first column contains the coin turned over by your friend.

Now for the rows. It is in the second row that things don't match up: there are an odd number of tails, but your "check digit" indicates that there should be an even number. Now you can read your friend's mind: "You turned over the coin in the first column, second row." A round of applause from the impressed audience.

What happens if your friend has turned over one of the coins you put down? No problem. Now the bottom right corner won't indicate the parity of either the last row or the last column. If it doesn't match the last row,

you'll know that one of the entries in the last row has been changed, so you
check each of the columns to see which one doesn't match up. If you find
that it's the sixth column that doesn't match, then it's actually the coin in
the bottom right corner that got turned.

Here's the grid again, but with one of your coins turned over. Can you
identify it?

H	H	T	T	T	H
T	T	H	T	T	H
H	H	H	T	H	T
T	H	H	T	T	T
T	T	T	T	T	T
T	H	H	T	H	H

Table 4.9

It's the one at the top right corner. The head at the bottom right
corner tells you that there should be an even number of tails sitting above
it in the last column—but there are an odd number. Now check the rows.
The first row doesn't match up because the head at the end of the row is
saying that there should be an even number of tails to the left of it. But
there are an odd number, and that tells you that the top right coin was
turned over.

This is the basis of what is called an error-correcting code, which is
used by computers to correct errors in messages that might have crept in
during transmission. Change the heads and tails into 0s and 1s, and suddenly, the grid becomes a digital message. For example, each column in the
5 × 5 grid that is laid down at the beginning of the trick could represent
a character in the Baudot code, and the grid in the trick would then be a
message five characters long. The extra columns and rows are added by the
computer to keep track of any errors.

So if we wanted to send the coded message on the front of Coldplay's
third album, we could use a similar trick applied to a 5 × 4 grid to detect
when and where any error occurred. Here is the album cover as it should
have been, with the colored blocks changed to 1s and the gaps to 0s:

1	1	0	1
0	1	1	0
1	0	0	1
1	1	1	0
1	1	1	1

Table 4.10

Now we add an extra column and row with 0s and 1s to indicate whether each column or row has an even or odd number of 1s:

1	1	0	1	1
0	1	1	0	0
1	0	0	1	0
1	1	1	0	1
1	1	1	1	0
0	0	1	1	0

Table 4.11

Now let's imagine that there's been an error during transmission and one of the numbers got flipped, and the graphic designer received the following message:

1	1	0	1	1
0	1	0	0	0
1	0	0	1	0
1	1	1	0	1
1	1	1	1	0
0	0	1	1	0

Table 4.12

By using the check digits in the last column and row, the graphic designer can spot the error. The second row and the third column don't match up.

Error-correcting codes like this are used in everything from CDs to satellite communications. You know what it's like when you're talking to someone on a phone and you can't make out everything that person says. When computers talk to each other, they have the same problem, but by using clever mathematics, we've managed to come up with ways to encode data so that we can get rid of this interference. This is what NASA did when the spacecraft *Voyager 2* sent back its first pictures of Saturn. By using an error-correcting code, NASA was able to turn fuzzy images into crystal-clear pictures.

HOW TO TOSS A COIN FAIRLY ACROSS THE INTERNET

Error-correcting codes help to communicate information clearly. But often, we want to use our computers to send secret information. In the past, Mary Queen of Scots, Lord Nelson, or anyone else intending to exchange secret messages needed to meet in advance with their agent to agree on the code that both parties would be using. In our modern computer age, we often need to send secret messages. When we're shopping online, we send our credit card details to people we've never met before, to websites we've just clicked on. Doing business on the Internet would be impossible with the old cryptography, where everyone first had to meet face-to-face. Luckily, mathematics provides a solution.

To explain the idea, let's start with a simple scenario. I want to play chess with someone over the Internet. I live in London, and my opponent is in Tokyo. We want to toss a coin to see who goes first. "Heads or tails?" I email my opponent. He writes back to say that he calls heads. I toss the coin. "Tails," I email back. "I start." Is there any way we can make sure I haven't cheated?

Amazingly, you can toss a coin fairly over the Internet, and it's the mathematics of prime numbers that makes it possible. All primes are odd except for 2 (which is the odd prime because it's the only even one). If we divide one of these odd prime numbers by 4, it leaves a remainder of 1 or 3. For example, 17 divided by 4 leaves a remainder of 1, and 23 divided by 4 has a remainder of 3.

As we know from chapter 1, the ancient Greeks proved two thousand years ago that there are an infinite number of primes. But are there an infinite number of primes that leave a remainder of 1 upon division by 4, or an infinite number that leave a remainder of 3? This was one of the questions that Pierre de Fermat challenged mathematicians with 350 years ago, though an answer had to wait until the nineteenth century and for the German mathematician Gustav Lejeune Dirichlet. He introduced some extraordinarily complicated mathematics to show that half the primes leave a remainder of 1 and half the primes leave a remainder of 3—no remainder is favored over the other. Just what mathematicians imply by "half" when they talk about the infinite is complicated. But essentially, it means that when you look at the primes that are less than any particular number, then pretty much half of them will have a remainder of 1 upon division by 4.

So whether a prime leaves a remainder of 1 or 3 upon division by 4 is no less "biased" than whether a fair coin lands heads or tails. For the purpose of our coin-tossing problem, let's associate heads with the primes with a remainder of 1 upon division by 4, and tails with primes that have a remainder of 3 upon division by 4. Now here comes the clever bit of math. If I take two prime numbers, say 17 and 41, both from the heads pile—the ones that have a remainder of 1 upon division by 4—and multiply them together, the answer also has a remainder of 1 upon division by 4: for example, $41 \times 17 = 697 = 174 \times 4 + 1$. If I take two primes, say 23 and 43, both from the tails pile and both with a remainder of 3 upon division by 4 . . . well, it's not what you might be expecting. When I multiply them together, they also give a number that has a remainder of 1 upon division by 4: in this case, $23 \times 43 = 989 = 247 \times 4 + 1$. So the product of the primes gives no hint about whether they're from the heads pile or the tails pile. This is what we can exploit to play "Internet heads or tails."

If I toss a coin and it lands heads, I choose two primes from the heads pile and multiply them together. If it lands tails, I choose two primes from the tails pile and multiply them together. Once I've tossed my coin and done my calculation, I send the answer to my opponent in Tokyo. It happens to be 6,497. Since the answer will always have a remainder of 1 upon division by 4, it's impossible for him, without knowing the primes, to tell whether the two primes I've chosen were from the heads pile or the tails pile. Now he is in a position to call heads or tails.

To see if he has won, I just need to send him the two primes I chose. In this case, they were 89 and 73—two primes from the heads pile. Since no other primes will multiply together to give 6,497, I have given him enough information with the number 6,497 to prove I haven't cheated, but not enough information so that he can cheat.

Actually, that's not strictly true. If he can crack 6,497 into 89 × 73, then he knows to call heads, but as long as I choose the primes big enough (much, much bigger than two-digit numbers), it's almost impossible with current computing power to crack the product into its prime factors. A similar principle is used in the codes that protect credit card numbers sent across the Internet.

An Easy Challenge

I've tossed a coin. I've taken two primes from the heads pile or from the tails pile and multiplied them together. The number I get is 13,068,221. Did the coin land heads or tails? Try answering this without a computer. (The answer is at the end of the chapter.)

A Tough Challenge

What if the number is 5,759,602,149,240,247,876,857, 994,004,081,295,363,338,151,725,852,938,901,132,472, 828,171,992,873,665,524,051,005,072,817,707,778,665, 601,229,693?

This time, you can use a computer.

WHY CRACKING NUMBERS
EQUALS CRACKING CODES

Bob runs a website selling soccer shirts in England. Alice lives in Sydney and wants to buy a shirt from the website, and she wants to send her credit card details without anyone else being able to see them. Bob publishes a special code number on his website; let's say it's 126,619. This code number is a bit like a key that locks away Alice's message and makes it secure. So when Alice visits the website, she gets a copy of the encoding key that Bob has published and uses it to "lock" her credit card.

What actually happens is that Alice's computer performs a special mathematical calculation on this number, 126,619, and her credit card number. The credit card number is now encoded and can be sent publicly over the Internet to Bob's website. (You can see the details of this calculation in the next section.)

But hold on—isn't there a problem with this? After all, if I'm a hacker, what's to stop me from visiting Bob's website, getting another copy of the key, and unlocking the message? The intriguing thing about these Internet codes is that you need a different key to unlock the door, and that key is kept very securely at Bob's headquarters.

The decoding key is the two primes that, multiplied together, give the number 126,619. What Bob actually does is choose the two primes 127 and 997 to build his encoding key, and it is these two primes that Bob has to use to undo the mathematical calculation that Alice's computer performed to scramble her credit card number in the first place. Bob has published the encoding key 126,619 on his website, but he keeps the two decoding primes 127 and 997 very secret.

If I can work out the two primes whose product is 126,619, I can hack into the card numbers being sent to Bob's website. Now, 126,619 is small enough for me to divide it by one number after another and find the primes 127 and 997 after not too long. You wouldn't be able to use this method on real websites, because their keys are based on much larger numbers—so large that finding the pair of primes by trial and error is almost impossible.

So confident are the mathematicians who invented the code that for many years, they were offering a $200,000 prize to anyone who could find the two prime factors of this 617-digit number:

25,195,908,475,657,893,494,027,183,240,048,398,571,429,282,126,
204,032,027,777,137,836,043,662,020,707,595,556,264,018,525,880,
784,406,918,290,641,249,515,082,189,298,559,149,176,184,502,808,
489,120,072,844,992,687,392,807,287,776,735,971,418,347,270,261,
896,375,014,971,824,691,165,077,613,379,859,095,700,097,330,459,
748,808,428,401,797,429,100,642,458,691,817,195,118,746,121,515,
172,654,632,282,216,869,987,549,182,422,433,637,259,085,141,865,
462,043,576,798,423,387,184,774,447,920,739,934,236,584,823,824,
281,198,163,815,010,674,810,451,660,377,306,056,201,619,676,256,
133,844,143,603,833,904,414,952,634,432,190,114,657,544,454,178,
424,020,924,616,515,723,350,778,707,749,817,125,772,467,962,926,
386,356,373,289,912,154,831,438,167,899,885,040,445,364,023,527,
381,951,378,636,564,391,212,010,397,122,822,120,720,357.

If you tried to crack this 617-digit number by trying one prime at a time, you'd need to work through more numbers than there are atoms in the universe before you got to them. Not surprisingly, the prize was never claimed, and in 2007, the offer was withdrawn.

As well as being virtually uncrackable, these prime-number codes have another rather novel feature, one which solved a problem that had dogged all previous codes. Before this prime-number code was invented, conventional codes were like a lock for which the same key is used to both lock and unlock the door. These Internet codes are like a new sort of lock: one key locks the door, but a different key unlocks it. This allows a website to freely distribute keys to lock messages, while keeping very secret the other key that unlocks messages. If you're feeling brave, here are the glorious details of how this Internet code really works. We start by introducing a curious calculator.

What Is a Clock Calculator?

The cutting-edge codes that are used on the Internet actually depend on a mathematical invention from hundreds of years before the Internet was ever dreamed of: the clock calculator. In the next section, we'll find out how clock calculators are used in the Internet codes, but first, we'll look at how these calculators work.

Let's start with the 12-hour clock. Addition on such a clock is something we're all familiar with—we know that four hours after nine o'clock will be one o'clock. This is the same as adding all the numbers together and working out the remainder after division by 12, and it is written like this:

$$4 + 9 = 1 \text{ (modulo 12)}$$

We write "modulo 12" because 12 is the modulus, the point after which the numbers start again. We could do similar sums on clocks with different numbers of hours rather than just sticking to 12. For example, on a clock with ten hours:

$$9 + 4 = 3 \text{ (modulo 10)}$$

How do we multiply numbers on a clock calculator? Multiplication consists of doing addition a certain number of times. For example, 4×9 means taking four 9s and adding them together. So where does the hand on the 12-hour clock end up after adding together four 9s? $9 + 9$ is the same as six o'clock. Each time we add another 9, the hand on the clock winds back three hours until we eventually get to 12 o'clock. Since 0 is such an important number in mathematics, we actually call this zero o'clock on a clock calculator. So we get this strange-looking answer:

$$4 \times 9 = 0 \text{ (modulo 12)}$$

What about raising a number to some power? Let's take 9^4, which means multiply 9 together four times. We've just learned how to do modular multiplication, so we should be able to do this easily enough. Because the numbers are getting quite big now, it will be easier here to take the remainder after division by 12 rather than trying to chase numbers around the clock. Let's start with 9×9, which is 81. What's the remainder upon division by 12—in other words, what is 81 o'clock? It turns out to be 9 again. And however many times we multiply 9 together, we always end up with 9:

$$9 \times 9 = 9 \times 9 \times 9 = 9 \times 9 \times 9 \times 9 = 9^4 = 9 \text{ (modulo 12)}$$

The answer on a clock calculator is achieved by calculating the answer on a normal calculator and then taking the remainder after division by the number of hours on the clock. But the strength of the clock calculator is that often you don't have to calculate things on the conventional calculator first. Can you work out what 7^{99} is on a 12-hour clock calculator? Hint: work out 7×7 first, then multiply the answer by 7 again. Do you see the pattern?

Fermat made a fundamental discovery about calculations on a clock calculator with a prime number of hours—say p—on it. He found that if you take a number on this calculator and raise it to the power p, you always get the number you started with. This is now called Fermat's little theorem, to distinguish it from his famous "last" theorem.

Here are some calculations on prime and nonprime clocks:

Power of 2	2^1	2^2	2^3	2^4	2^5	2^6	2^7	2^8	2^9	2^{10}
On a conventional calculator	2	4	8	16	32	64	128	256	512	1,024
On a 5-hour clock calculator	2	4	3	1	2	4	3	1	2	4
On a 6-hour clock calculator	2	4	2	4	2	4	2	4	2	4

Table 4.13

Now, 5 is prime, and if I calculate 2 to the power of 5 on a five-hour clock calculator, the answer is 2 again. So $2^5 = 2$ (modulo 5). The magic is guaranteed to work if the number of hours on the clock is prime. It doesn't necessarily work if you take a non-prime-number clock. For example, 6 isn't prime, which explains why on the six-hour clock, 2^6 ends up not at 2 but at 4.

As the clock hand maps out the hours, a pattern emerges. After $p - 1$ steps, we are guaranteed at the next step to return to where we started, so the pattern repeats itself every $p - 1$ steps. Sometimes the pattern repeats itself several times during the $p - 1$ steps. On a clock with 13 hours, here's what we see when we run through the different powers of 3 from 3^1, 3^2, and so on up to 3^{13}: 3, 9, 1, 3, 9, 1, 3, 9, 1, 3, 9, 1, 3.

The hand doesn't visit all the times on the clock, but there is still this repeating pattern that brings it back to three o'clock after multiplying 3 together 13 times.

We've already seen similar math at work in chapter 3, in the perfect shuffles that can be used to cheat at poker. There, we varied the number of cards in the pack and asked how many perfect shuffles it would take to return the pack to its original order. A pack with $2N$ cards can sometimes take a full $2N - 2$ shuffles, but sometimes it can take far fewer. For a pack of 52 cards, only eight perfect shuffles sees the pack return to its original order, while a pack with 54 cards needs 52 perfect shuffles.

Fermat never fully explained his working-out, and he left it as a challenge to future generations of mathematicians to explain his discovery and show why it always works for prime-number clocks. It was Leonhard Euler who eventually came up with a proof for why this magic always works on prime-number clocks.

Fermat's Little Theorem

Here's an explanation of Fermat's little theorem. The theorem says that, on a clock with a prime p *number of hours on it,*

$$A^p = A \ (modulo \ p)$$

The proof is tough, but not technical: you'll just need to stay focused to follow it.

Let's start with an easy case to get us going. If A = 0, the theorem is true because however many times we multiply 0 together, we always get 0. So let's suppose that A is not 0. We'll set out to show that if we multiply A together p − 1 times on this clock, then we get to one o'clock. This will suffice to prove the theorem, because multiplying 1 by A again must bring us to A.

First, we make a list of all the hours on the clock, excluding 0. There are p − 1 of them: 1, 2, . . . , p − 1. Now we multiply each number in the list by A on our clock calculator and get

$$A \times 1, A \times 2, \ldots, A \times (p − 1) \ (modulo \ p)$$

The hours in this list must be just the same as those in the original list—1, 2, . . . , p − 1—but in a different order; if this were not the case, then either one of the answers is 0 or two answers are the same. There isn't room for anything else to happen, because there are only p hours on the clock.

Suppose that A × n and A × m give the same time on the p-hour clock, where n and m lie between 1 and p − 1 (I'll show why this means that n = m). So A × n − A × m = A × (n − m) is 0 on the clock calculator—that is, A × (n − m) on our ordinary calculator is divisible by p.

The key to the next step of the proof is to use the fact that p is a prime number. Like a chemical molecule, the number A × (n − m) is built by multiplying together the prime-number atoms that make up A and the prime-number atoms that make up n − m. Now, p is prime—one of the atoms of arithmetic—and can't be cracked any further. Because p divides into A × (n − m) with no remainder, it must be one of the atoms used to build A × (n − m), because there is only one way to build a number by multiplying primes. But p doesn't divide into A with no remainder, so it must be in the list of atoms used to build n − m. In other words, n − m is divisible by p. But what does that mean? It means that n and m *are the same time on our p-hour clock. You can use a similar argument to show that* A × n *can't be zero o'clock if neither* A *nor* n *are zero o'clock.*

Note that it is very important that the clock have a prime number of hours—we have seen already that 4 × 9 *is 0 on the 12-hour clock without either 4 or 9 being 0.*

Now we have two lists—1, 2, . . . , p − 1 and A × 1, A × 2, . . . , A × (p − 1)—*consisting of the same numbers but in a different order. Here we can use a nice trick, one that Fermat himself had probably discovered. If we multiply all the numbers on either list together, we get the same answer, because the order in which we multiply things doesn't matter. The first list gives us* 1 × 2 × . . . × (p − 1), *which we can write as* (p − 1)! *The second list consists of* A *multiplied together p − 1 times, and again 1 to p − 1 multiplied together. After a little rearrangement, this gives us* $(p − 1)! × A^{p−1}$. *And these give the same answer on our clock calculator:*

$$(p − 1)! = (p − 1)! × Ap^{−1} \ (modulo \ p)$$

This means that $(p − 1)! × (1 − A^{p−1})$ *is divisible by p, and we use the same trick as before. None of the numbers 1, 2, . . . , p − 1 are divisible by p, so* (p − 1)! *can't be divisible by p. The only possibility is that* $1 − A^{p−1}$ *is divisible by p. And this means that the calculation* $A^{p−1}$ *on the clock calculator always gives the answer 1—just what Fermat had teased mathematicians to explain.*

There are several interesting ingredients in this argument. It is certainly important that if A × B *is divisible by a prime p, then either* A *or* B *must also be divisible by that prime—something that follows from the special property of the primes. But the beautiful moment for me comes in seeing the same thing—this list of numbers 1, 2, . . . , p − 1—in two different ways: lateral thinking at its best.*

How to Use a Clock to Send Secret Messages over the Internet

We are now almost ready to show how these clocks are used to send secret messages over the Internet.

When you buy something on a website, your credit card number is encrypted by your computer using the website's public clock calculator, so the website needs to tell your computer how many hours are on this clock. This is the first of two numbers that your computer receives. Let's call this number N. In our example of Bob's soccer-shirt website, this number was 126,619. There is also a second encoding number that your computer needs to do the calculation, which we'll call E. Your credit card number C is encoded by raising it to the power E, as calculated by the clock calculator with N hours. In this way, you get the scrambled number C^E (modulo N), and that's the number your computer sends to the website.

But how does the website unscramble this number? Fermat's prime-number magic is the key. Let's suppose that N is a prime-number clock. (We'll see later that this isn't quite good enough for a secure code, but it will help us understand where we're going.) If we multiply the number C^E together enough times, then C will magically reappear. But how many times (D) must we multiply C^E together? In other words, when does $(C^E)^D = C$ on the clock with p hours?

Of course, if $E \times D = p$, this works. But p is prime—so there can't be any such number D. Now, if we keep multiplying C together, then there is another point in which we are guaranteed to get C appearing again as the answer. The next time the credit card number appears is when we raise it to the power $2(p - 1) + 1$. It appears again when we raise it to the power $3(p - 1) + 1$. So to find the decoding number, we need to find a D such that $E \times D = 1$ (modulo $(p - 1)$). This is a much easier equation to solve. The problem is that because E and p are public numbers, it's also easy for a hacker to find the decoding number D. To make it safe, we must use a discovery made by Euler about clocks with $p \times q$ hours on them, not just p hours.

If you take a time C on a clock with $p \times q$ hours, how long does it take for $C, C \times C, C \times C \times C, \ldots$ to repeat itself? Euler discovered that the pattern repeats itself after $(p - 1) \times (q - 1)$ steps. So to get back to the original time, you need to raise C to the power $(p - 1) \times (q - 1) + 1$, or $k \times (p - 1) \times (q - 1) + 1$, where k is the number of times the pattern repeats itself.

So now we know that in order to decode a message CE on a clock with $p \times q$ hours, we must find a decoding number D such that $E \times D = 1$ (modulo $(p - 1) \times (q - 1)$), so we have to do a calculation on a secret clock calculator with $(p - 1) \times (q - 1)$ hours. A hacker knows only the numbers N and E, and if he wants to find the secret clock, he must uncover the secret primes p and q. Cracking an Internet code is therefore the same as cracking a number N into its prime building blocks. And as we saw in the section on flipping a coin over the Internet, this is virtually impossible when the numbers are big.

Let's look at the Internet code in action, but with very small p and q so that we can easily follow what's going on. We'll say that for his soccer-shirt website, Bob has chosen the primes 3 and 11, so that the public clock calculator that customers must use to encrypt their credit card number has 33 hours on it. Bob keeps the primes 3 and 11 secret because they are the key to decoding messages, though he makes the number 33 public because this is the number of hours on his public clock calculator. The second piece of information made public on Bob's website is the encoding number E—let's say it's 7. Everyone who buys a shirt from Bob online is doing exactly the same thing: raising their credit card number to the power 7 on a 33-hour clock calculator.

Bob's soccer-shirt website is visited by a customer who was one of the first ever recipients of a credit card, and its number is 2. Raise 2 to the power 7 on the 33-hour clock calculator, and you get 29.

Here's a clever way to calculate 2^7 on the 33-hour clock calculator. We need to start multiplying 2s together: $2^2 = 4$, $2^3 = 8$, $2^4 = 16$, $2^5 = 32$. As we go to higher powers of 2, the hand on the clock winds farther around the clock face, and when we get to multiply by 2 for a sixth time, the hand is doing more than a whole revolution. There's a little trick we can do here that makes it look as though the clock hand is reversing its direction rather than spinning around farther. We simply say that 32 o'clock on our 33-hour clock calculator is −1 o'clock. Then, after we've gotten to $2^5 = 32$, two more multiplications by 2 will get us to −4, or 29 o'clock. This way, we are spared having to calculate 2 to the power 7— that is, 128—and then finding the remainder upon dividing by 33. For very big numbers, this sort of saving is invaluable for a computer trying to calculate things quickly.

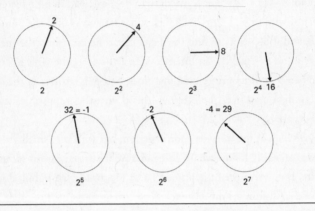

Figure 4.17   Calculating powers on a 33-hour clock calculator.

How can we be sure that the customer's encrypted number, 29, is secure? After all, a hacker can see this number travelling through cyberspace and can easily look up Bob's public key, consisting of the clock calculator with 33 hours and the instruction to raise the credit card number to the power 7. To crack this code, all the hacker needs to do is find a number that, multiplied together 7 times on the 33-hour clock calculator, gives the answer 29.

Needless to say, it's not that easy. Even with normal arithmetic, squaring numbers can be done on the back of an envelope, but it's much tougher to undo the process and search for the square root. The extra twist comes from computing powers on a clock calculator. You very quickly lose sight of the starting point because the size of the answer bears no relationship to where you started.

In our example, the numbers are small enough for the hacker to be able to try every variation to find the answer. In practice, websites use clocks in which the number of hours has over one hundred digits, so an exhaustive search is impossible. You may well be wondering how, if it is so difficult to solve this problem on the 33-hour clock calculator, any Internet trading company can recover a customer's credit card number.

Euler's more general version of Fermat's little theorem guarantees the existence of a magic decoding number, D. Bob can multiply the encrypted credit card number together D times to reveal the original credit card number. But you can work out what D is only if you know the secret primes p and q. Knowledge of these two primes becomes the key to unlocking the secrets of this Internet code, because you must solve the following problem on the secret clock calculator:

$$E \times D = 1 \text{ (modulo } (p-1) \times (q-1))$$

When we put in our numbers, we find that we have to solve the equation

$$7 \times D = 1 \text{ (modulo } (2 \times 10))$$

That's asking us to find the number that, when multiplied by 7, gives a number with a remainder of 1 upon division by 20. $D = 3$ will work because $7 \times 3 = 21 = 1$ (modulo 20).

And if we raise our encrypted credit card number to the power 3, the original card number reappears:

$$29^3 = 2 \text{ (modulo } 33)$$

Being able to recover the credit card number from the encoded message depends on knowing the secret primes p and q, so anyone wanting to hack the codes on the Internet needs a way to take a number N and crack it into its prime divisors. Every time you buy a book online or download a music track, you are using the magic of prime numbers to keep your credit card secure.

THE MILLION-DOLLAR QUESTION

The code makers are always trying to keep ahead of the code breakers. In case the prime-number code is ever cracked, mathematicians are constantly coming up with ever more clever ways to send secret messages. A new code called elliptic curve cryptography, or ECC for short, is already being used to protect the flight paths of aircraft, and the million-dollar prize for this chapter relates to understanding the mathematics of the elliptic curves behind these new codes.

There are a multitude of different elliptic curves, but they all have equations like $y^2 = x^3 + ax + b$. Each curve corresponds to different values of a and b: for example, $a = 0$ and $b = -2$ gives us $y^2 = x^3 - 2$.

This equation defines a curve that I can draw on graph paper, as shown in the following figure, by finding a succession of points (x, y). I put in a value of x, and then I calculate the equation $x^3 - 2$ and take its square root to get the corresponding value of y. For example, if $x = 3$, then $x^3 - 2 = 27 - 2 = 25$. To get y, I need to take the square root of 25, since $y^2 = x^3 - 2$, so y is 5 or -5 (because minus times minus is plus, there are always two square roots). The graph you get is symmetrical about the horizontal axis because all the square roots have a mirror root that is negative. Here, we've found the two points $(3, 5)$ and $(3, -5)$.

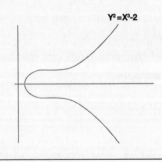

$Y^2 = X^3 - 2$

Figure 4.18   The graph of an elliptic curve.

Those points on the elliptic curve are very nice because x and y are both whole numbers. Can you find any other points like this? Let's try putting $x = 2$. Then $x^3 - 2 = 8 - 2 = 6$, so $y = \sqrt{6}$ or $-\sqrt{6}$. In the first example, 25 had a whole number square root, but the square root of 6 is not so tidy. The ancient Greeks proved that there is no fraction, let alone whole number, that when you square it it gives 6. $\sqrt{6}$ written as a decimal number races off to infinity with no pattern at all:

$\sqrt{6} = 2.449489742783178\ldots$

The million-dollar question relates to finding the points on this curve where both x and y are whole numbers or fractions. Most of the time they aren't, because when you put in x, the y won't be a whole number or even a fraction because most numbers don't have a nice square root. We were lucky to find that $(3, 5)$ and $(3, -5)$ are nice points on the curve, but are there any others?

The ancient Greeks came up with a beautiful piece of geometry that showed how to get more points (x, y), where both x and y are fractions, once you've found one. Draw a line that just touches the first point you've found—it mustn't go through it; it must be at just the right angle to glance the curve, as shown in the following graph. We call this line the tangent to the curve at that point. By extending this line, we find that it will cut the curve at a new point. The exciting discovery is that the coordinates of this new point will also both be fractions.

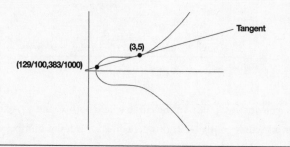

Figure 4.19   How to find more points on the elliptic curve with coordinates that are fractions.

For example, if we draw the tangent at the point $(x, y) = (3, 5)$ on the elliptic curve $y^2 = x^3 - 2$, we find that it crosses the curve at a new point,

$(x, y) = (^{129}/_{100}, {}^{383}/_{1,000})$, where both coordinates are fractions. With this new point, we could repeat the procedure and get another point where both x and y are fractions:

$$\left(\frac{2,340,922,881}{45,427,600}, \frac{93,955,726,337,279}{306,182,024,000} \right)$$

Without this bit of geometry, it would be very hard to discover that feeding in the fraction

$$x = \frac{2,340,922,881}{45,427,600}$$

will give you a y that is also a fraction.

In this example, you can keep repeating this bit of geometry and get an infinite number of pairs of fractions (x, y) that are points on this curve. For a general elliptic curve $y^2 = x^3 + ax + b$, if you've got one point $(x1, y1)$ on the curve where both $x1$ and $y1$ are fractions, then setting

$$x_2 = \frac{(3x_1^2 + a)^2 - 8x_1 y_1^2}{4y_1^2}$$

and

$$y_2 = \frac{x_1^6 + 5ax_1^4 + 20bx_1^2 - 5a^2 x_1^2 - 4abx_1 - a^2 - 8b^2}{8y_1^2}$$

will give you another point on the curve where both $x2$ and $y2$ are fractions.

For our curve $y^2 = x^3 - 2$, this generates an infinite number of points on the curve where both x and y are fractions, but there are curves where it's impossible to get an infinite number of points. For example, take the curve defined by the equation $y^2 = x^3 - 43x + 166$. On this curve, it turns out that there are only a finite number of points where both x and y are whole numbers or fractions:

$(x, y) = (0, 0), (3, 8), (3, -8), (-5, 16), (-5, -16), (11, 32), (11, -32)$

In fact, they all have whole-number coordinates. Trying to use the geometry trick or algebra to get more points with fractions just delivers one of these seven points again.

The million-dollar question, called the Birch and Swinnerton-Dyer conjecture, asks whether there is a way to tell which elliptic curves will have an infinite number of points where both coordinates are whole numbers or fractions.

You might say who cares? Well, we all should, because the mathematics of elliptic curves is now being used in mobile phones and smart cards to protect our secrets, as well as in air traffic control systems to ensure our safety. With this new code, your credit card number or message is converted by clever math into a point on this curve. To encrypt the message, the math moves the point around to another point using the geometry just explained to generate new points.

Undoing this geometric procedure requires some math that currently we can't do. But if you manage it and crack this million-dollar problem, you probably wouldn't worry about the million dollars, because you'd end up being the most powerful hacker on the planet.

SOLUTIONS

Substitution Cipher Decoded

"A mathematician, like a painter or a poet, is a maker of patterns.

"If his patterns are more permanent than theirs, it is because they are made with ideas. The mathematician's patterns, like the painter's or the poet's, must be beautiful; the ideas like the colours or the words, must fit together in a harmonious way. Beauty is the first test: there is no permanent place in the world for ugly mathematics."

The cipher is the following:

Plaintext	a	b	c	d	e	f	g	h	i	j	k	l	m
Ciphertext	B	A	N	T	S	H	U	F	L	K	X	I	O

Plaintext	n	o	p	q	r	s	t	u	v	w	x	y	z
Ciphertext	C	M	Q	P	V	E	D	G	R	Z	W	J	Y

Table 4.14

An Easy Challenge

It landed heads. 13,068,221 = 3,613 × 3,617. Both 3,613 and 3,617 are
primes that have a remainder of 1 upon division by 4. There is a way to
factorize this number quickly, using a method discovered by Fermat. If you
square 3,615, you get 13,068,225, which is 4 away from 13,068,221. 4 is
also a square. Now, if you use a bit of algebra that tells you that $a^2 - b^2 =$
$(a + b) \times (a - b)$, you get

$$13,068,221 = 3,615^2 - 2^2 = (3,615 + 2) \times (3,615 - 2) = 3,613 \times 3,617.$$

Five

THE QUEST TO PREDICT THE FUTURE

I f time travel were possible, it would be easy to predict the future—I could just come back from next year and tell you what happens. Sadly, though, we don't yet know how to travel through time, and many of the ways in which people claim to be able to predict the future, such as gazing into crystal balls or casting horoscopes, are complete mumbo jumbo. If you really want to know what's going to happen tomorrow, next year, or far into the next millennium, your best bet is mathematics.

Mathematics can predict whether the earth will be hit by an asteroid and how long the sun will keep burning. But there are still some things that even mathematicians find hard to forecast. We have the equations to explain, for example, the weather, population growth, and the turbulence behind a soccer ball moving through the air, but some of those equations we don't know how to solve. The million-dollar prize for this chapter will go to the person who can crack the equations of turbulence and predict what will happen next.

The ability of mathematics to predict the future has given those who understand the language of numbers immense power. From the astronomers of ancient times who could predict the movements of the planets in the night sky to the hedge fund managers of today who predict movements of prices on the stock market, people have used mathematics to take a peek into the future. The power of mathematics was recognized by St. Augustine, who warned, "Beware of mathematicians, and all those who make empty prophecies. The danger already exists that the mathematicians have made a covenant with the Devil to darken the spirit and to confine man in the bonds of Hell."

While some modern math is admittedly devilishly hard, rather than keeping us in the dark, its practitioners are constantly on the lookout for new ideas to shed light on future events.

HOW DID MATH SAVE TINTIN?

In Hergé's comic-strip book *Prisoners of the Sun,* the young Belgian reporter Tintin is taken prisoner by an Inca tribe after he strays inside the Temple of the Sun God. The Incas condemn Tintin and his friends Captain Haddock and Professor Calculus to be burned at the stake. The fire is to be lit by a magnifying glass that will focus the sun's rays onto the pile of wood. Tintin is, however, allowed to choose the time of their death. But can he use this favor to save himself and his friends?

Tintin does the math and learns that a solar eclipse will hit the area in a few days' time, so he chooses the time of their death to coincide with the eclipse. (Actually, someone else did the math; he saw the prediction in a newspaper cutting.) Shortly before the eclipse is due, Tintin calls out, "The Sun God will not hear your prayers! O magnificent Sun, if it is thy will that we should live, give us now a sign!" And just as the math predicted, the sun disappears and the terrified tribe releases Tintin and his friends.

Mathematics is the science of spotting patterns, and that's how math gives us the power to look into the future. Early astronomers gazing at the night sky soon realized that the movements of the moon, sun, and planets repeat themselves. Many cultures use these celestial patterns as a way of keeping track of the passing of time. Many different calendars are possible because the sun and moon dance to a crazy syncopated rhythm as they make their way across the sky, but one thing these calendars have in common is the role of mathematics in making sense of the cycles of the moon and the sun to mark time. What's curious is the role played by the number 19 in determining when moving holidays like Easter are celebrated each year.

The basic unit of time common to all these calendars is the 24-hour day. This isn't the time it takes the earth to spin once on its axis—that's a little less: 23 hours, 56 minutes, and 4 seconds. If we were to use this slightly shorter period as the length of a day, our clock and the rotating earth would get more and more out of sync as all those extra 3 minutes and 56 seconds accumulated, until eventually, noon on the clock would happen at midnight. So for timekeeping purposes, we define a day—or, to use the correct term, a solar day—as the time it takes the sun to return to the same position in the sky as seen from the same point on the earth's surface. After one complete rotation, the earth will have moved around its orbit by about $1/365$ of a revolution, so it takes about $1/365$ of a rotation, or another $1/365$ of a day—about 3 minutes and 56 seconds—for the sun to return to the same point in the sky.

To be a little more precise, the earth takes 365.2422 of these solar days to go once around the sun. The Gregorian calendar, the one used by most countries, is based on a close approximation to this cycle. 0.2422 is almost a quarter, so by adding an extra day to the calendar every four years, the Gregorian calendar keeps in pretty good step with the earth's movement in its orbit around the sun. A few tweaks are required because 0.2422 isn't quite 0.25: every one hundred years, we miss out on a leap year, and every four hundred years, we skip the skipping and retain the leap year.

The Islamic calendar uses the cycle of the moon instead. Here, the basic unit is a lunar month, and 12 of these make up a lunar year. A lunar month, the beginning of which is determined by the sighting of the new moon at Mecca, is about 29.53 days, making a lunar year 11 days shorter than a solar year. 365 divided by 11 is approximately 33, so it takes 33 years for the month of Ramadan to cycle its way through the solar year, which is why Ramadan slips through the year as reckoned by the Gregorian calendar.

The Jewish and Chinese calendars mix and match, using the cycle of the earth's orbit around the sun and the cycle of the moon's orbit around the earth. They do this by adding a leap month roughly every third year, and the key to the calculations is the magic number 19. Nineteen solar years (= 19 × 365.2422 days) almost exactly matches 235 lunar months (= 235 × 29.53 days). The Chinese calendar has seven leap years in every 19-year cycle to keep the lunar and solar calendars in sync.

The number 19 would have been important in Tintin's calculations because the sequence of eclipses of the sun and moon also repeats itself after 19 years. This episode in *Prisoners of the Sun* is based on a famous moment in history, when the explorer Christopher Columbus used a lunar eclipse, rather than a solar eclipse, to save his crew when they were stranded on Jamaica in 1503. The local inhabitants were friendly at first but eventually became hostile and refused to supply Columbus and his crew with provisions. With his men facing starvation, Columbus came up with a cunning plan. He consulted his almanac—a book that contained predictions of tides, lunar cycles, and positions of stars used by sailors for navigation—and discovered that a lunar eclipse was due on February 29, 1504. Columbus summoned the locals three days before the event and threatened them: if they didn't supply him with provisions, he would make the moon disappear.

No supplies turned up—the locals didn't believe that Columbus had the power to make the moon disappear. But on the evening of February 29, when the moon rose above the horizon, they could see that a piece had already been bitten out of it. According to Columbus's son Ferdinand, as the moon faded from the night sky, the natives became terrified, and "with great howling and lamentation came running from every direction to the ships laden with provisions, praying to the Admiral to intercede with his god on their behalf." By precise calculation, Columbus timed his pardon of the locals to coincide with the moon's gradual reappearance. This story might be apocryphal or embellished by the Spanish to contrast the clever European conquerors with the ignorant locals. But at its heart, it shows the power of mathematics.

When Will the Next Eclipse Be?

If you know the time of one eclipse, you can use a mathematical equation to work out the time of another one. The calculations depend on two important numbers.

The first is the synodic month (S) of 29.5306 days. This is the average time it takes for the moon to go around the earth and return to the same position relative to the sun. It's the average time between two new moons.

The other is the draconic month (D) of 27.2122 days. The moon's orbit around the earth is slightly tilted with respect to the earth's orbit around the sun. The two orbits cross each other in two places, called the nodes of the moon's orbit, as shown in the following figure. The draconic month is the average time it takes the moon, starting from one node, to pass through the opposite node and return to the first one.

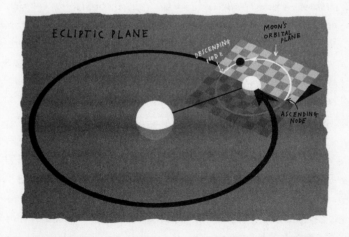

Figure 5.1   *The orbit of the moon intersects the orbit of the earth in two places, called the ascending node and the descending node.*

For every pair of whole numbers A and B you can find that will make A × S very close to B × D, you will have the date of an eclipse A × S ≈ B × D days after the last eclipse you saw. And there will be another eclipse after another A × S ≈ B × D days. The sequence of eclipses will continue for a while, but because the equation is not exact, the eclipses will eventually get less and less impressive, until the sun, moon, and earth are no longer aligned. And that will be the end of that particular cycle of eclipses.

Here's an example: A = 223 synodic months is very close to B = 242 draconic months, so every 223 × 29.5306 ≈ 242 × 27.2122 days after an eclipse, there will be another almost identical eclipse. That is a period of approximately 6,585⅓ days, or about 18 years, 11 days, and 8 hours. The shift of 8 hours means that the next two such eclipses will be seen from a different location on the surface of the earth. However, the third will hit the same spot, so every three times 18 years, 11 days, and 8 hours—or approximately 19,756 full days—there will be a repeat eclipse.

For example, the total lunar eclipse visible from North America on December 21, 2010, is a repeat of the eclipse on December 9, 1992, seen from Europe. It was last seen in America on November 18, 1956. There have been other eclipses between these dates, but they are part of other eclipse cycles that are running alongside this one. The math helps you calculate the date of the next eclipse in each cycle.

The power of mathematics to predict happenings in the night sky relies on spotting patterns that repeat themselves. But how can we predict something new? The story of how we can use the equations of mathematics to look into the future begins with predicting the behavior of simple objects, such as a soccer ball.

IF I DROP A FEATHER AND A SOCCER BALL, WHICH HITS THE GROUND FIRST?

The soccer ball, of course. You don't have to be a world-class mathematician to predict that. But what if I drop two soccer balls of the same diameter, one containing lead and the other air? For most people, their first reaction is to say that the lead ball will hit the ground first. That was certainly the belief of Aristotle, one of the greatest thinkers of all time.

In an apocryphal experiment, the Italian scientist Galileo Galilei showed that this intuitive answer is completely wrong. He worked in Pisa, home to the world-famous Leaning Tower. Where better to chuck things over the side and have your apprentice standing at the bottom to see which lands first? Galileo proved Aristotle wrong: both balls, though of different weights, will hit the ground simultaneously.

Galileo realized that the weight of the object didn't come into it. What makes a feather fall more slowly than a ball is air resistance, and if you could take the air away, a feather and a ball should both fall at the same speed. One place where you could test this theory is on the airless surface of the moon. In 1971, the commander of the Apollo 15 mission to the moon, David Scott, recreated Galileo's experiment by dropping a geological hammer and a falcon's feather at the same time. They fell much more slowly than they would on earth because of the moon's lower gravitational pull, but the two objects hit the ground simultaneously, just as Galileo predicted they would.

NASA's lunar rerun of Galileo's experiment can be viewed at http://nssdc.gsfc. nasa.gov/planetary/lunar/apollo_15_ feather_drop.html or by using your smartphone to scan this code.

As the mission controller said later, the result was "reassuring considering both the number of viewers that witnessed the experiment and the fact that the homeward journey was based critically on the validity of the particular theory being tested." That's very true: space travel would be impossible to plan without having the equations of mathematics to predict a spacecraft's flight as it is pushed and pulled by the gravity of the earth, sun, moon, and planets and the thrust of its engines.

Once he had discovered that the weight of a falling object was irrelevant to its speed, Galileo wanted to see whether he could predict how long it would take for the object to hit the ground. Objects fell too quickly for accurate timing from something like the top of the Leaning Tower, so Galileo decided to roll balls down a slope to see how the speed varied. He discovered that if a ball rolled one unit of distance after one second, then after two seconds it would have covered four units of distance, and after three seconds it would have traveled nine units. He could then predict that after four seconds the ball should have traveled a total of 16 units of distance—in other words, the distance a body falls is proportional to the square of the time for which it has been falling. In mathematical symbols, $d = \frac{1}{2}gt^2$ where d is the distance fallen and t is the time. The factor g, known as the acceleration due to gravity, told Galileo how the vertical speed of a falling object changed with each passing second. For a soccer ball dropped from the top of the Leaning Tower of Pisa, after one second its speed is g, after two seconds it is $2g$, and so on. Galileo's formula was one of the first examples of a mathematical equation being used to describe nature, of what would come to be called a law of physics.

Using mathematics in this way has revolutionized the way we understand the world. Previously, people had used everyday language to describe nature, and that can be vague—you could say that something was falling, but you couldn't say when it would land. With the language of mathematics, people could not only describe nature more precisely, but they could also predict the way it would behave in the future.

Having worked out what happens to a ball when you drop it, Galileo's next move was to predict what happens to it when it gets kicked.

WHY DOES WAYNE ROONEY SOLVE A QUADRATIC EQUATION EVERY TIME HE VOLLEYS THE BALL INTO THE GOAL?

"Beckham with the free-kick, a perfectly timed volley by Rooney . . . goal!!!"

But how did Rooney do it? You might not think so, but Rooney has to be incredibly good at math to be able to score such a goal. Every time he gets on the end of a free kick from Beckham, he's subconsciously working out another of the equations Galileo concocted so that he can predict where the ball will end up.

Equations are like recipes. Take the ingredients, mix them up a certain way, and the equation spits out an outcome. To construct the equation that Rooney will solve, Galileo needs the following ingredients: the incoming ball's horizontal speed u and vertical speed v when it left Beckham's foot, and the effect of gravity, which is summed up in the number g and tells Rooney how the vertical speed of the soccer ball changes with each second. The value of g depends on what planet you play your soccer; on the earth, gravity increases the speed by about 9.8 meters per second per second (about 22 mph per second). Galileo's equation then tells Rooney the height of the ball at any point relative to where the free kick was taken. For example, if the ball is a horizontal distance of x meters from where Beckham kicked the ball, then the height above the ground will be y meters, where y is given by the equation

$$y = \frac{v}{u}x - \frac{g}{2u^2}x^2$$

The recipe is the set of mathematical instructions for what to do with all these numbers, and the outcome is the height of the ball at a certain point in its trajectory.

For Rooney to work out how far to stand from the free kick so that he can volley or head the ball into the net, he has to undo the equation and work backward. First, he decides that he wants to head the ball. Rooney is about 1.80 meters tall, so the ball has to be at a height $y = 1.80$ if he is going to head it (without jumping). He knows what u, v, and g are. Let's choose some approximate numbers: $u = 20$, $v = 10$, and $g = 10$. (If you are worried about units, the speeds u and v are in meter/second, and g is in meter/second^2.)

The only thing Rooney doesn't know is how far away from Beckham he should stand to intercept the ball correctly. But the equation does have that information encoded in it; it's just not so visible. The equation says that Rooney should stand x meters from Beckham, where x is a number that makes the equation

$$1.8 = \frac{10}{20}x - \frac{10}{2 \times 400}x^2$$

true. Tidying this up gives us

$$x^2 - 40x + 144 = 0.$$

This sort of equation might look familiar—it's one we all learned to solve in school; it's called a quadratic equation. Think of it as a cryptic crossword clue, hiding the true value of x.

Amazingly, the first people to start solving equations like this one were the ancient Babylonians. Their quadratic equations didn't describe trajectories of soccer balls but appeared when they were surveying the land around the Euphrates. A quadratic equation comes about when we are trying to work out some quantity that is multiplied by itself. We call this squaring because it gives the area of a square, and it's in the context of calculating the area of a piece of land that these quadratic equations were first formulated.

Here's a typical problem. If a rectangular field has an area of 55 square units and one side is six units shorter than the other, how long is the longer side? If we call the longer side x, then the problem tells us that $x \times (x - 6) = 55$ or, simplifying things,

$$x^2 - 6x - 55 = 0.$$

But how do you go about unraveling this mathematical cryptic clue?

The Babylonians came up with a neat method: they dissected the rectangle and rearranged the pieces to make a square, which is an easier shape to deal with. We can divide up the pieces of our field just as Babylonian scribes would have done thousands of years ago:

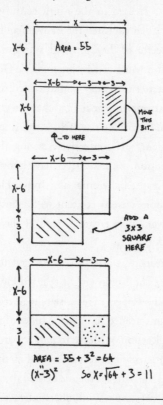

Figure 5.2   How to solve a quadratic equation by completing a square.

Start by cutting a small rectangle measuring 3 x (x – 6) units off the end of the rectangle and move this around to the bottom of the rectangle. The total area hasn't changed—just the shape. The new shape is almost a square with each side x – 3 units long, but it's missing a small 3 x 3 square in the corner. If we add in this small square, we increase the area of the shape by nine units. The area of this large square is therefore 55 + 9 = 64. Now we have the simple task of taking the square root of 64 to discover the length of the side, which must be 8. But the side had length $x - 3$, and so $x - 3$ = 8—that is, $x = 11$. Although we've only been shuffling around imaginary parcels of land, behind what we've been doing lies a general method for unlocking those cryptic quadratics.

Once algebra was created in the ninth century in Iraq, a formula could be written down that captured the Babylonian method. Algebra was developed by the director of the House of Wisdom in Baghdad, a man named Muhammad ibn-Musa al-Khwarizmi. The House of Wisdom was the top intellectual center of its day and attracted scholars from around the world to study astronomy, medicine, chemistry, zoology, geography, alchemy, astrology, and mathematics. The Muslim scholars collected and translated many ancient texts, effectively saving them for posterity— without their intervention, we may never have known about the ancient cultures of Greece, Egypt, Babylonia, and India. However, the scholars at the House of Wisdom weren't content with translating other people's mathematics. They wanted to create a math of their own and push the subject forward.

Intellectual curiosity was actively encouraged in the early centuries of the Islamic empire. The Koran taught that worldly knowledge brought people closer to holy knowledge. In fact, Islam required mathematical skills, because devout Muslims needed to have the times of prayer calculated and needed to know the direction of Mecca to pray toward.

The algebra of al-Khwarizmi revolutionized mathematics. Algebra is a language that explains the patterns that lie behind the behavior of numbers, and its grammar underlies the way numbers interact. A bit like a code for running a program, it will work with whatever numbers you feed into the program. Although the ancient Babylonians had devised a cunning method to solve particular quadratic equations, it was al-Khwarizmi's algebraic formulation that ultimately led to a formula that could be used to solve any quadratic equation.

Whenever you have a quadratic equation $ax^2 + bx + c = 0$, where a, b, and c are numbers, then the geometric juggling can be translated into a formula with x on one side and a recipe combining the numbers a, b, and c on the other:

$$x = \frac{-b + \sqrt{b^2 - 4ac}}{2a}$$

It is this formula that allows Rooney to undo the equation controlling the flight of the ball and work out how far away to stand. We left him knowing that he had to stand x meters from the position of the free kick, where

$$x^2 - 40x + 144 = 0$$

Using algebra, he can work out that he should stand 36 meters away from Beckham to intercept the ball with his head.

How did he do this? Well, in the quadratic equation controlling Beckham's free kick, $a = 1$, $b = -40$, and $c = 144$. So the formula for undoing this equation tells us that the distance Rooney should stand from Beckham is

$$x = \frac{40 + \sqrt{1,600 - 4 \times 144}}{2} = 20 + \frac{\sqrt{1,024}}{2} = 20 + 16 = 36 \text{ metres}$$

Interestingly, because -32 is also the square root of 1,024, we get another solution: $x = 4$ meters. This is the point where the ball is heading upward on its trajectory; Rooney will wait until the ball starts coming down again. Because there is always a negative as well as a positive square root, we always get two solutions out of this formula. To indicate this, sometimes the equation comes with a +− instead of a + in front of the square root symbol.

Of course, Rooney is using a much more intuitive approach—one that doesn't require him to do mental math for 90 minutes. But it does show how the human brain is almost programmed by evolution to be good at making predictions.

WHY DOES A BOOMERANG COME BACK?

Strange things happen to objects when they spin. When you kick a soccer ball off-center, it bends in the air, and when you toss a tennis racket into the air, it always spins before you catch it. A spinning gyroscope seems to defy gravity by leaning horizontally. But the classic example of the strange behavior of a spinning object is the way a boomerang comes back.

The dynamics of spinning objects are very complicated and have foxed generations of scientists. But we now know that the reason a boomerang comes back has to do with two different factors. The first relates to the lift of an airplane's wing, and the second is called the gyroscopic effect. Mathematical equations help to explain and ultimately predict how the geometry of a wing generates a force that pushes up, counteracting the force of gravity that is pulling the airplane down. An airplane's wings are shaped so that the air flows faster over the top of the wing than underneath it. The air on top gets squashed and pushed faster over the wing. It's the same principle as water flowing through a pipe: where the pipe narrows, the water flows faster.

A second equation, called the Bernoulli equation, shows that the higher speed of the air over the top of the wing leads to a lower pressure above the wing, and the lower speed of the air below the wing leads to a higher pressure. This difference in pressures above and below the wing creates force, which lifts the airplane.

If you look at a classic boomerang closely, you can see that each arm is shaped like the wing of an airplane. This is what makes the boomerang turn. To throw a boomerang with the hope that it will return, you need to launch it from a vertical position in such a way that (thinking of it like an airplane) its right wing is at the top and its left wing is at the bottom. The same force that lifts an airplane's wing now pushes the boomerang to the left.

But there's something slightly more subtle happening here. If the boomerang were simply behaving like an airplane, then the force would just send it to the left, not make it come back. It comes back because when it's thrown, it's given spin, which, thanks to the gyroscopic effect, causes the force pushing it to the left to constantly change direction, with the result that the boomerang is pushed around the arc of a circle.

When I throw the boomerang, the top part is spinning forward and the bottom part is spinning backward. The upper part is like an airplane's wing, traveling faster through the air. On an airplane flying horizontally, that faster movement would create more lift. But on a boomerang being launched vertically, it causes the boomerang to tilt, its top leaning into the eventual arc of its flight.

Now the gyroscopic effect comes into play. When you place a spinning gyroscope on a stand in a vertical position, it behaves like a top. But if you tilt it so that its axis of rotation is at an angle to the vertical, something called precession happens: the axis of rotation itself begins to rotate. This is what happens with the spinning boomerang. Its axis of rotation is an imaginary line running through its center, and as this axis rotates, so the boomerang is pushed around a circle.

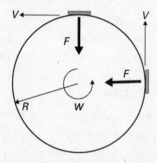

Figure 5.3   The forces acting on a boomerang. F is the force created by the lift, V is the speed at which the center of the boomerang moves, R is the radius of the path of the boomerang, and W is the rate of precession.

Anyone who has thrown a boomerang will know that getting it to come back isn't easy. You need to throw the boomerang in such a way that *V,* the speed at which it leaves your hand, and *S,* the spin rate you give the boomerang when you launch it, satisfy the formula

$$a \times S = \sqrt{2}V$$

where *a* is the radius of the boomerang—the distance from its center to its tip. By flicking your wrist more, you can increase *S* in an attempt to get the formula to work.

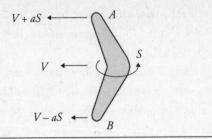

Figure 5.4   The top of the boomerang, A, travels faster than the bottom, B, thanks to the spin.

The angle at which the boomerang tilts depends on the difference in the forward speed of the top and bottom of the boomerang. The top is traveling at speed $V + aS$ while the bottom is going slower, at speed $V - aS$, where S is the angular speed measuring the rate at which the boomerang is spinning around its center (see figure 5.5). You can therefore vary the tilt of the boomerang by changing the speeds V and S, which will have an effect on how quickly the boomerang precesses, or twists, as it moves around its circular arc at speed V. If your boomerang won't come back, it may be that you aren't flicking it at the right speed S in relation to the launch speed V. This equation will help you adjust your throw accordingly.

Once you've mastered how to get the boomerang to come back to you, will throwing it harder and faster make it turn in an even bigger arc? The mathematics can be pieced together to give an equation that tells us the radius of the boomerang's circular trajectory. Once again, the equation is a recipe that takes various ingredients that define the boomerang and its flight, mixes them together, and outputs the radius. Here are the ingredients:

- J, the moment of inertia of the boomerang. This is a measure of how hard it is to spin the boomerang; the heavier the boomerang, the greater J is. The moment of inertia also depends on the boomerang's shape.
- ρ, the density of the air through which the boomerang flies

- C_L, the lift coefficient—a number that determines the amount of lift the boomerang experiences, depending on its shape
- π, the number 3.14159 . . .
- a, the radius of the boomerang

The radius of the boomerang's path R is determined by mixing these ingredients together using the following recipe:

$$R = \frac{4J}{\rho C_L \pi a^4}$$

With this equation, we see that if we throw the boomerang harder and faster, it doesn't change the radius because speed isn't one of the ingredients in the recipe. But what happens if we make the boomerang heavier by sticking a bit of Blu-Tack at the end of each wing? The equation helps us to predict that an increase in weight will increase the moment of inertia J, and that will increase the radius R. So a heavier boomerang goes around in a bigger circle—useful to know before throwing boomerangs around in a confined space!

You can download a PDF from the Number Mysteries website with instructions to make your own boomerang.

Can You Make an Egg Defy Gravity?

Take a hard-boiled egg. Lay it on its side on a table and give it a spin. Miraculously, the egg stands up, seemingly defying the laws of gravity. Even more strangely, if you try it with a raw egg, the same magic doesn't happen.

It took until 2002 for mathematicians to come up with an explanation for this behavior. The rotational energy is translated, via the friction on the table, into potential energy, which pushes the egg's center of gravity upward. If the table has no friction or too much friction, then the effect doesn't happen. Part of the energy transferred to a raw egg gets absorbed by the fluid interior, and there isn't enough left to push the egg upright.

WHY ARE PENDULUMS NOT AS PREDICTABLE AS THEY FIRST APPEAR?

It was Galileo, the master of using math to make predictions, who first unlocked the secret of what makes a pendulum tick. The story goes that when he was 17, he was attending Mass at the cathedral in Pisa. In a moment of boredom, he stared up at the ceiling, and his eyes fell on a chandelier that was swinging gently in the breeze blowing through the building.

Galileo decided to time how long it took the chandelier to swing from side to side. He didn't have a watch (they hadn't been invented yet), so he used his pulse to keep track of the swing. The great discovery he made was that the time the chandelier took to complete one swing did not seem to depend on the size of the swing. In other words, the time of the swing essentially doesn't change if you increase or decrease the angle of swing. (I put the word *essentially* in there to indicate that if we dig a little deeper, things get slightly more complicated.) When the wind blew harder, the chandelier swung through a larger arc but took the same time to swing as when the wind dropped and the chandelier was hardly moving at all.

This was an important discovery and resulted in the swinging pendulum being used to record the passage of time. If you are starting a pendulum clock, you don't have to worry about how far to the side you are going to move the pendulum, especially as the angle of swing will decrease over time. But what does the time of the swing depend on, and can we predict whether and how the swing will change if the weight is increased or if the pendulum is made longer?

As we might guess from Galileo's experiment at the Leaning Tower of Pisa, a heavier pendulum does not travel faster, so a pendulum's swing does not depend on its weight. But increasing the length of the pendulum does have an effect on the time of the swing. It turns out that multiplying the length by 4 doubles the time. Multiply the length by 9 and the time triples; multiply it by 16 and the time quadruples.

Again, an equation can capture this prediction. The time of swing T goes up in step with the square root of the length L:

$$T \approx 2\pi \sqrt{\frac{L}{g}}$$

This is actually another way of writing the equation that Galileo created for dropping balls from the Leaning Tower: the g is, again, the acceleration due to gravity. The reason for the $\approx$ as opposed to an $=$ and my earlier use of *essentially* is that this is a good approximation for the time a pendulum takes to swing from one side to the other. As long as the swing isn't too big, it's possible to use this equation to predict the pendulum's behavior. But if the angle of swing is large—if we start the pendulum almost vertically, for example—then the math gets much more difficult. Now the angle starts to have an effect on the time of swing, which Galileo didn't pick up because the chandelier in the cathedral couldn't swing that far. We also don't see the effect in a grandfather clock, because the pendulum's swing is quite small.

The mathematics you need to find the right equation to predict the behavior of a pendulum with a large angle of swing goes beyond what is taught in most math degree courses. Here is the beginning of the formula. It actually has an infinite number of parts that contribute to the behavior of the pendulum. θ_0 is the initial angle the pendulum makes with a vertical line.

$$T \approx 2\pi \sqrt{\frac{L}{g}} \left(1 + \frac{1}{16}\theta_0^2 + \frac{11}{3{,}072}\theta_0^4 + \ldots \right)$$

But this is nothing compared with the problem of predicting the behavior of a slightly modified pendulum. Instead of a single rigid rod swinging to and fro, imagine adding a second pendulum hinged to the bottom of the first, so that the whole thing looks a little like a leg, with an upper and lower part hinged at the knee. Predicting the behavior of this double pendulum is extremely complex. It's not that the equations are that much more complicated, but that their solutions are very unpredictable: by changing the initial position of the pendulum very slightly, the result can be dramatically different. This is because the double pendulum is an example of a mathematical phenomenon called chaos. But a double pendulum isn't just an amusing desktop game. The math behind it has important consequences for a question that could affect the future of humanity itself.

One of the many computer simulations of the double pendulums found online can be viewed at www.myphysicslab.com/ dbl_pendulum.html. You can also watch it by using your smartphone to scan this code.

Try to predict whether the lower piece will next pass clockwise or counterclockwise through the top pendulum. It is almost impossible.

To make your own pendulum, see www.instructables.com/id/The-Chaos-Machine-Double-Pendulum or scan this code with your smartphone.

WILL THE SOLAR SYSTEM FLY APART?

Since Galileo first investigated falling balls and swinging pendulums, mathematicians have formulated hundreds of thousands of equations that predict how nature behaves. These equations are the foundations of modern science and are known as laws of nature. Math has enabled us to create the complex technological world in which we live. Engineers rely on equations to reassure them that bridges won't fall down and airplanes will stay in the air. You might think from our story so far that predicting the future would always be easy, but it's not always that simple—as the French mathematician Henri Poincaré discovered.

In 1885, King Oscar II of Sweden and Norway offered a prize of twenty-five hundred kroner for anyone who could establish mathematically once and for all whether the solar system would continue turning like clockwork, or whether it was possible that at some point, the earth might spiral away from the sun and off into space. Poincaré thought he could find the answer, and he began to investigate.

One of the classic moves that mathematicians make when they are analyzing complicated problems is to simplify the setup in the hope that it will make the problem easier to solve. Instead of starting with all the planets in the solar system, Poincaré began by considering a system with just two bodies. Isaac Newton had already proved that their orbits would be stable: the two bodies just travel in elliptical orbits around each other, forever repeating the same pattern.

Figure 5.5

From this starting point, Poincaré began to explore what happens when another planet is added into the equation. The problem is that as soon as you have three bodies in a system, for example the earth, moon, and sun, the question of whether their orbits are stable gets very complicated—so much so that it had stumped even the great Newton. The problem is that now there are 18 different ingredients to combine in the recipe: the exact coordinates of each body in each of three dimensions, and their velocities in each dimension. Newton himself wrote that "to consider simultaneously all these causes of motion and to define these motions by exact laws admitting of easy calculation exceeds, if I am not mistaken, the force of any human mind."

Poincaré was not daunted. He made significant headway, simplifying the problem by making successive approximations to the orbits. He believed that if he rounded up or down the very small differences he found in the positions of the planets, it wouldn't affect his final answer too much. Although he couldn't solve the problem in its entirety, his ideas were sophisticated enough to win him King Oscar's prize. However, when Poincaré's paper was being prepared for publication, one of the editors couldn't follow Poincaré's mathematics and raised a question. Could Poincaré justify why making a small change in the positions of the planets would result in only a small change in their predicted orbits?

While Poincaré was trying to justify his assumption, he suddenly realized he'd made a mistake. Contrary to what he'd originally thought, even a small change in the initial conditions—the starting positions and velocities of the three bodies—could end up producing vastly different orbits: his simplification didn't work. He contacted the editors and tried to stop the paper from being printed, because publishing an erroneous paper in honor of the king would have caused a furor. The paper had already been printed, but most of the copies were gathered together and destroyed.

It all seemed a huge embarrassment. But as often happens in mathematics, when something goes wrong, the reason it goes wrong leads to interesting discoveries. Poincaré wrote a second, extended paper explaining his belief that very small changes could cause an apparently stable system to suddenly fly apart. What he discovered through his mistake led to one of the most important mathematical concepts of the last century: chaos theory.

Poincaré had found that even in Newton's clockwork universe, simple equations can produce extraordinarily complex results. This isn't the mathematics of randomness or probability. We are dealing here with a system that mathematicians call deterministic: it's controlled by strict mathematical equations, and for any set of starting conditions, the outcome will always be the same. A chaotic system is still deterministic, but a very small change in the starting conditions can lead to a vastly different outcome.

Here is a small-scale example that acts as a good model of the solar system. We put three magnets—one black, one gray, and one white—on the floor. Above the magnets, we set up a magnetic pendulum that is free to swing in any direction. The pendulum will be attracted by all three magnets and will swing among them before assuming a stable position. At the end of the pendulum is a cartridge that drips out a trail of paint. We set the pendulum swinging, and the dripping paint traces out the path of the pendulum. What we are really attempting to simulate is an asteroid whizzing through a solar system with three planets attracting it: eventually, the asteroid will hit one of the planets.

The extraordinary thing is that it is almost impossible to repeat the experiment and get the same paint trail. As hard as you try to set the pendulum off from the same position and in the same direction, you find that the paint traces out a completely different path and ends up attracted to a different magnet each time. Here are three paths that start in almost the same way but end up at different magnets:

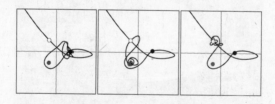

Figure 5.6   Just a small change in the initial position of the pendulum can cause it to follow a completely different path between the three magnets (shown as the small white, gray, and black dots).

The equations controlling the path of the pendulum are chaotic, and just a very small change in the starting location has a dramatic effect on the outcome. This is the signature of chaos.

We can use a computer to produce a picture of which magnet the pendulum will be attracted to. The magnets are located at the centers of the corresponding three large vase-shaped blocks of color. If you start the pendulum over a black region, it will eventually settle over the black magnet.

Similarly, if you start the pendulum over a gray or white region, it will end up at the gray or white magnet. You can see regions of this picture in which moving the pendulum's starting point a little won't affect the outcome dramatically. For example, if you start near the black magnet, the pendulum is likely to end its journey at the black magnet. But there are other regions where the colors change rapidly over small distances.

This is an example of a shape that nature likes very much—the fractal. Fractals are the geometry of chaos, and if you zoomed in on some of the regions of this picture, you would see the same level of complexity, as we found on page 85. It is this complexity that makes the motion of the pendulum so hard to predict, even though the equations describing it are quite simple.

Figure 5.7 This computer-generated image illustrates the behavior of the pendulum moving over the magnets.

What if it's not just the outcome of a swinging pendulum, but the future of the solar system that's at stake? Perhaps a small perturbation by a rogue asteroid will cause a change that's small but enough to send the solar system spiraling apart. This seems to have happened in the nearby solar system of the star Upsilon Andromedae, where astronomers believe that the strange behavior of the existing planets is evidence of a catastrophe in which one of the original planets orbiting the star was ejected after something disturbed the previously stable orbits. Could the same thing happen to our planet?

Just to reassure themselves, scientists have recently used supercomputers to try to answer the question that ultimately defeated Poincaré: is the earth really in danger of flying off into space? They ran the actual orbits of the planets backward and forward in time. Fortunately, the calculations showed that, with a 99 percent probability, the planets will continue to run smoothly in their orbits for another five billion years (by which time the sun will have evolved into a red giant star and swallowed up the inner solar system). But that still leaves a 1 percent chance of an outcome that's slightly more interesting—at least mathematically.

It turns out that the rocky inner planets (Mercury, Venus, Earth, and Mars) have less stable orbits than the gas giants (Jupiter, Saturn, Uranus, and Neptune). Left to their own devices, these big planets would have a remarkably stable future. It's tiny Mercury that has the potential to cause the solar system to go into catastrophic meltdown.

Computer simulations reveal that a strange resonance between Mercury and Jupiter might evolve, which could cause Mercury's orbit to start crossing the orbit of its nearest neighbor, Venus. That would set the stage for a potential almighty collision between Venus and Mercury, which would probably rip the solar system apart. But will this really happen? We don't know. Chaos makes predicting the future very hard.

HOW CAN A BUTTERFLY KILL THOUSANDS OF PEOPLE?

It's not just the solar system that's chaotic. Many natural phenomena show chaotic traits: the behavior of the stock market, the buildup of a freak wave at sea, the beating of the heart. But the chaotic system that impacts everyone's lives the most is the weather. "Will the earth still be orbiting the sun in a billion years' time?" is not such an immediate concern. We want to know whether it will be warm and sunny next week, and ultimately whether the climate in 20 years' time is going be dramatically different from its current state.

Weather forecasting has always been something of a dark art, even though some of the folklore to do with the weather has now been proved true. "Red sky at night, shepherd's delight" works because the sun's rays have been reddened by travelling through a large region of clear skies to the west of the shepherd. As weather systems in Europe generally arrive from the west, this signals good weather on the way.

Today's meteorologists have a multitude of data to work with, ranging from measurements by weather stations at sea to images and information from satellites. And they have immensely precise equations to describe how the clashing air masses in the atmosphere interact to create clouds, wind, and rainfall. If we have the mathematical equations that control the weather, then can't we just run the equations with today's weather data through a computer and see what it will be like next week?

Alas, even with today's supercomputers, a forecast for the weather two weeks in advance is still not reliable. We can't know precisely what today's weather will be, let alone what it will be further ahead. Even the best weather stations have only a limited accuracy. We can never know the exact speed of every particle in the air, the precise temperature at every point in space, or the exact pressure all across the planet—and just a small variation in any of these could produce a wildly different weather forecast. This has given rise to the phrase "the butterfly effect": a butterfly flapping its wings makes tiny changes in the atmosphere that just possibly could ultimately cause a tornado or a hurricane to form on the other side of the world, wreaking havoc, taking lives, and causing millions of dollars of damage.

For this reason, meteorologists run several weather predictions simultaneously, each starting with a slight variation in the measurements taken from the worldwide networks of weather stations and satellites. Sometimes, all the predictions come up with broadly similar results, and the meteorologists can then be fairly confident that the weather—though technically chaotic—will be stable for the next week or two. But on some runs, the predictions differ completely, and the forecasters know there's no way they can accurately predict the weather even days ahead.

With our chaotic pendulum swinging between the three magnets, there were regions in the picture predicting the pendulum's behavior where a small change in the initial position of the pendulum wouldn't have caused it to end up at a different magnet. And it's the same with the weather. Think of the large black region in Figure 5.7 as the weather in a desert: it's always going to be hot there, however hard a butterfly flaps its wings—and similarly for the arctic, which is like the magnet staying in a white region. But the weather for the United Kingdom is like the pendulum starting at a place where the colors in the picture change rapidly with just a small shift in the pendulum's position.

If we knew the precise positions and speeds of all the particles in the universe, we could predict the future with certainty. The problem is that if you get one of those starting positions even slightly wrong, the future can turn out to be completely different. The universe may behave like clockwork, but we'll never know the positions of the cogs accurately enough to take advantage of its deterministic nature.

HEADS OR TAILS?

The 1968 European Football Championship was held before penalties were introduced as a way of deciding a drawn match. So when Italy and the Soviet Union were still goalless after extra time in their semifinal, a coin was tossed to decide which of them would go through to the final. It has been universally acknowledged since Roman times that a coin is a fair way to decide a dispute. After all, it's impossible to tell as it spins through the air how it will land. Or is it?

Theoretically, if you knew precisely the position of the coin, how much it was being spun, and when it would land, you could calculate how the coin would land. But like the weather, wouldn't a minute change in any one of these factors potentially cause a completely different outcome? Persi

Diaconis, a mathematician at Stanford University, California, decided to test whether coin tossing is as unpredictable as we think. If the conditions are the same whenever you toss the coin, then the math will always produce the same outcome. But is the signature of chaos hidden inside the toss of a coin? What if we change those starting conditions very slightly—do the changes get amplified, so that by the time the coin lands it's impossible to know whether it will be heads or tails?

With help from his engineering friends, Diaconis built a mechanical coin-tossing machine that could replicate the conditions of the coin toss over and over again. Of course, there would be very minor differences from one toss to the next, but would these differences result in wildly different outcomes, like the pendulum swinging between the three magnets? Diaconis found that every time he repeated the experiment with his mechanical coin-tosser, the coin would always land the same way. He then trained himself to be able to flip the coin in an identical way each time so that he could get ten heads in a row. Make sure not to gamble on the toss of a coin with the likes of Persi Diaconis.

But what about average human tossers, who will change the way they launch a coin from one flip to the next? Diaconis wondered whether there might still be some bias. To begin his mathematical analysis, he needed an expert in spinning objects. He knew he had his man when he met Richard Montgomery, whose claim to fame was proving the falling-cat theorem—a theory that explains why a cat dropped from any angle always manages to land on its feet. Together with statistician Susan Holmes, they showed that a spinning coin launched with a flick of the thumb has a bias toward landing with a particular side up.

To convert the theory into actual numbers, they needed to do some careful analysis of how a spinning coin moves through the air. With the help of a high-speed digital camera that shoots ten thousand frames per second, they captured the motion of a coin and fed the data into their theoretical model. What they found may come as a surprise: there is indeed a bias in a true toss of the coin. It's small: 51 percent of the time, the coin tended to land with the same face upright as when it was spun into the air. The reason seems to relate to the physics of the boomerang or gyroscope. It appears that the spinning coin also precesses like a gyroscope, and so spends slightly more time in the air with the face that was first showing pointing upward. The difference is inconsequential for one throw, but in the long run, it could be very significant.

One organization that definitely cares about the long run is the casino. Their profit depends on long-term probabilities. For every throw of the dice or spin of the roulette wheel, they rely on your failing to predict how the dice or ball will land. But just as with the tossed coin, if you knew precisely the starting positions of the roulette wheel and the ball, and their starting speeds, you could in theory apply Newtonian physics to determine where the ball would land. Start the roulette wheel in exactly the same position and with exactly the same speed, and have the croupier launch the ball in exactly the same way each time, and the ball will land in exactly the same place. The problem here is the same one that Poincaré discovered: even a very small change in the starting positions and speeds of the roulette wheel and ball can have a dramatic effect on where the ball ends up. And it's the same with dice.

But that doesn't mean that mathematics can't help you narrow down where the ball might end up. You can watch the ball spinning around the wheel a few times before you place your bet, so you have the chance to analyze the trajectory of the ball and predict its final destination. Three Eastern Europeans—a Hungarian woman described as "chic and beautiful" and two "elegant" Serbian men—did just that. They used mathematics to make a killing at the roulette wheel at the London Ritz casino in March 2004.

Using a laser scanner hidden inside a mobile phone linked to a computer, they recorded the spin of the roulette wheel relative to the ball over two rotations. The computer worked out a region of six numbers within which it predicted the ball would fall. During the third rotation of the wheel, the gamblers placed their bets. Having reduced their odds of winning from 37:1 to 6:1, the trio placed bets on all six numbers in the section where the ball was predicted to finish. That first night, they netted £100,000. On the second night, they won a staggering £1.2 million. Despite being arrested and kept on police bail for nine months, they were eventually released and allowed to keep their winnings. Legal teams concluded that they had done nothing to tamper with the wheel.

The gamblers realized that although there is chaos in a roulette wheel, a small change in the starting positions of the ball and wheel don't always lead to huge changes in the outcome. This is what meteorologists rely on when they are predicting the weather. Sometimes when they run their computer models, they find that changing the conditions of the weather today doesn't have a dramatic effect on the forecast. The gamblers' computer was doing the same thing, running through thousands of different scenarios to see where the ball might end up. It couldn't identify the position precisely, but a region of six numbers was enough to turn the odds in the gamblers' favor.

You might think from what you've read so far that nature divides itself into problems that are simple and predictable (like a ball falling from the top of the Leaning Tower of Pisa) and problems that are chaotic and hard to predict (like the weather). But it's not so clear-cut. Sometimes what starts out as easily predictable can become chaotic if one small thing changes by just a fraction.

WHO KILLED ALL THE LEMMINGS?

Some years ago, environmentalists noticed that every four years, the numbers of lemmings seem to plummet dramatically. A popular theory was that every few seasons, these Arctic rodents made their way to a tall cliff and threw themselves over the edge, plunging to their deaths on the rocks below. In 1958, Walt Disney Productions' natural history unit included footage of this mass suicide in its award-winning film *White Wilderness*. The sequence looked so convincing that the word *lemming* came to be used for anyone who unquestioningly follows the masses with potentially disastrous consequences. The animals' behavior even spawned a video game in which players had to save the lemmings from their mindless march toward the cliff's edge.

To see footage from White Wilderness, *go to www.youtube.com/watch?v=xMZlr 5Gf9yY or use your smartphone to scan this code.*

In the 1980s, it was revealed that the film crew of *White Wilderness* had faked the whole sequence. According to a Canadian TV documentary, the lemmings, which had been bought specially for the filming, refused to leap over the cliff on cue—so members of the film crew "encouraged" them over the edge. But if it isn't mass suicide that's causing these sudden drops in the number of lemmings every four years, then what is the explanation?

Figure 5.8

Yet again, it turns out that mathematics has the answer. A simple equation tells us how many lemmings there will be from one season to the next. We start by assuming that, because of environmental factors such as food supply and predators, there's a maximum population that can be sustained. We'll call that N. We'll say that L is the number of lemmings that survived from the previous season, and that after the births in the new season, the population rises to K lemmings. A proportion of these K lemmings will not survive. The proportion that dies is L/N; namely, the number of lemmings in the previous season divided by the maximum population possible. So $K \times L/N$ die, leaving

$$K - \frac{K \times L}{N}$$

lemmings at the end of this season. To make our calculations simple, let's say that the maximum population is $N = 100$.

This equation, although simple, has some surprising outcomes. Let's start by looking at what happens if the lemming population doubles each spring, so that $K = 2L$. Of these, $2L \times L/100$ will die. Let's suppose that in the first season, there are 30 lemmings. The equation predicts that at the end of the second season, there will be $60 - (60 \times 30/100) = 42$ lemmings. They carry on increasing in number until by the fourth season, there are 50 of them.

From then onward, the number of lemmings that survive each season remains constant at 50. The surprising thing is that, whatever the original population at the start of the first season, the number of lemmings left at the end of each subsequent season will always eventually home in on half the maximum number, where it will remain. So once you hit 50 lemmings, the number doubles to 100 during the next season, but by the end of the season, $100 \times 50/100 = 50$ will have died, leaving a population of 50 lemmings again.

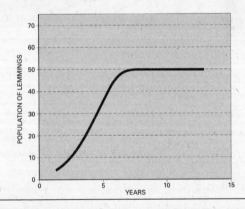

Figure 5.9

If lemmings double their numbers each spring, the population reaches a stable value however many lemmings there are to start with.

What happens if the lemmings are more productive? If the number of lemmings slightly more than triples from one season to the next, the population doesn't stabilize but instead flips between two values. In one season, the number of lemmings that survive to the end of the season is quite high; in the following season, it falls.

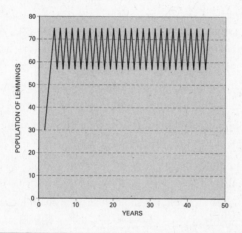

Figure 5.10   If the lemmings triple their numbers in the spring, the population starts to oscillate.

When the lemmings become even more productive, the population starts to fluctuate in a strange way. If the population increases by a factor of 3.5, then the total number of lemmings oscillates between four values, repeating this pattern every four seasons. (The precise factor at which four values first appear is $1 + \sqrt{6}$, which is approximately 3.449.) And this is where we find that in one of these four years there can be a significant drop in the number of lemmings, not because of a mass suicide pact but because of the math.

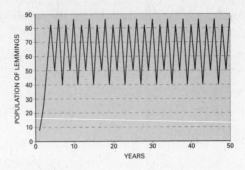

Figure 5.11   When lemming numbers increase in the spring by a factor of 3.5, the population oscillates between four different values.

The really interesting change in the population dynamics happens when the lemmings increase their numbers by a factor of just over 3.5699. Then, their numbers from one season to the next jump around seemingly without any rhyme or reason. Even though the equation that calculates the population is a simple one, it has started to produce chaotic results. Change the initial number of lemmings, and the population dynamics are completely different. Beyond the threshold where the chaos kicks in, 3.5699, it's almost impossible to predict how the population will vary. The equation controlling the population numbers can start out very predictable, but with just a small change in lemming fecundity, chaos can suddenly erupt.

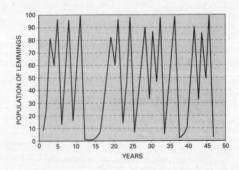

Figure 5.12   When lemming numbers increase in the spring by a factor of 3.5699 or more, the population variations become chaotic.

How to Play the Fishy Formula Game

This is a game for two players. Download the PDF from the Number Mysteries website and cut out the ten fish and the fish tank. The game explores how the number of fish varies over ten seasons. Each fish corresponds to one season, and there is a box on the side of the fish in which you can keep track of the number of fish in the tank corresponding to that season. The fish tank can sustain a maximum of 12 fish. The fish survive for one year, and during that year, they have a certain number of offspring and then die.

Roll two dice. The number of fish that start in the tank is then the number on the dice minus one (so it is a number between 1 and 11). Call this number N^0. The first player chooses a number, K, between 1 and 50. This will determine how many offspring each fish has. If there are N^0 fish in the tank to begin with, then during the first year, they give birth to $(^K\!/_{10}) \times N^0$ fish. The number of fish is therefore multiplied by $^K\!/_{10}$, a number between 0.1 and 5.

Not all the new fish survive. If there were N fish in the tank at the end of the previous year, then by the end of the next year, the number of fish will be

$$\frac{K}{10} \times N \times \left(1 - \frac{N}{12}\right)$$

You must round up or down to get a whole number of fish (4.5 fish is rounded up to 5).

Let the fish tank "run" for ten years. The first player scores the number of fish in the tank at the end of odd-numbered years, and the second player scores the number of fish in the tank at the end of even-numbered years.

Let N^i be the number of fish in year i. So

player 1 scores $N^1 + N^3 + N^5 + N^7 + N^9$,

player 2 scores $N^2 + N^4 + N^6 + N^8 + N^{10}$.

By writing on the fish you cut out, you can keep track of the population numbers from one year to the next. If all the fish die at some point, then player 1, who chose the multiplier K, loses automatically.

Here's an example. The players throw the two dice and score a 4. So there are three fish in the tank at the beginning: $N_0 = 3$. Player 1 chooses K = 20. The number of fish at the end of year 1 is therefore

$$N_1 = \frac{K}{10} \times N_0 \times \left(1 - \frac{N_0}{12}\right) = 2 \times 3 \times \left(1 - \frac{3}{12}\right) = 4.5 \approx 5$$

In year 2, there are

$$N_2 = \frac{K}{10} \times N_1 \times \left(1 - \frac{N_1}{12}\right) = 2 \times 5 \times \left(1 - \frac{5}{12}\right) = 5\frac{5}{6} \approx 6$$

fish, and in year 3, there are

$$N_2 = \frac{K}{10} \times N_2 \times \left(1 - \frac{N_2}{12}\right) = 2 \times 6 \times \left(1 - \frac{6}{12}\right) = 6$$

fish. The number of fish has now stabilized because 6 gets repeated when it is put into the formula. So

player 1 scores 5 + 6 + 6 + 6 + 6 = 29 fish,

player 2 scores 6 + 6 + 6 + 6 + 6 = 30 fish,

and player 2 wins. See what happens when you vary the multiplier K.

Because we are rounding numbers up and down, the game doesn't have the full subtlety of the chaotic model that killed off the lemmings.

For an online tank simulator to accompany this game, visit http://www.rigb. org/christmaslectures06/50_20.html or use your smartphone to scan this code.

In this version of the game, the number of fish displayed has been rounded up or down, but the fraction of fish is fed into the formula for the next year. For example, if you set K = 27 and N0 = 3,

N_1 = 6.075, rounded to 6 fish;

N_2 = 8.09873, rounded to 8 fish;

N_3 = 7.10895, rounded to 7 fish;

N_4 = 7.8233, rounded to 8 fish;

N_5 = 7.352, rounded to 7 fish;

N_6 = 7.68872, rounded to 8 fish;

N_7 = 7.45835, rounded to 7 fish;

N_8 = 7.62147, rounded to 8 fish;

N_9 = 7.50844, rounded to 8 fish;

N_{10} = 7.58804, rounded to 8 fish.

Player 1 scores 6 + 7 + 7 + 7 + 8 = 35 fish,

player 2 scores 8 + 8 + 8 + 8 + 8 = 40 fish.

HOW TO BEND IT LIKE BECKHAM
OR CURL IT LIKE CARLOS

David Beckham and Roberto Carlos have hit some extraordinary free kicks in their soccer careers, kicks that seem to defy the laws of physics. Perhaps the most amazing was the one Carlos took for Brazil against France in 1997. The free kick was awarded 30 meters from the goal. Most soccer players would just have kicked the ball to a teammate and gotten play under way again. Not Roberto Carlos. He placed the ball on the ground and stepped back, ready to shoot.

The French goalkeeper, Fabien Barthez, lined up a defensive wall, though he couldn't really have believed that Carlos was going to aim the ball anywhere near his goal. And sure enough, when Carlos ran up and struck the ball, it appeared to be heading well wide of its mark. Spectators to one side of the goal started ducking, expecting the ball to fly into the crowd. Then suddenly, at the last minute, the ball veered to the left and flew into the back of the French goal. Barthez couldn't believe what he'd just seen. He hadn't moved an inch. "How on earth did the ball do that?" you could see him thinking.

*You can see footage of Roberto Car-
los's free kick at www.youtube.com/
watch?v=Pb2qykj6_ZU&feature=fvst,
or use your smartphone to scan this code.*

Carlos's strike, far from defying physics, took advantage of the science of moving soccer balls. The effect of spin on a soccer ball can cause it to do some crazy things. Kick a ball without giving it any spin, and it will travel as if it's moving across a fixed two-dimensional sheet of paper, tracing out a parabola. But put some spin on the ball, and suddenly the mathematics of its motion becomes three-dimensional. As well as moving up and down, it can also swing left or right.

So what is pushing the ball to the left or right as it flies through the air? It's a force called the Magnus effect, named after the German mathematician Heinrich Magnus, who in 1852 was the first to explain the effect of spin on a ball. (The Germans have always been good at soccer.) It's similar to how lift on an airplane's wing is created. As I explained on page 222, the different speeds of the air flowing over and under the wing cause a decrease in pressure above the wing and a higher pressure below it, producing a force that pushes the wing up.

To get the ball to swing from right to left, Carlos kicked it so that its left side spun toward him (spinning around a vertical axis through the center of the ball). The ball's spin was then effectively helping to push the air past it more quickly on the left side. That caused the air on the left side to travel faster past the ball, decreasing the pressure—the same as happens on the top of an airplane's wing. The pressure on the right side of the ball went up because the air speed there was decreasing as the surface of the ball was spun into the path of the air flowing past it. The increased pressure translated into a force pushing the ball from right to left, which eventually took the ball into the back of the net.

The same principle is used to make a golf ball travel farther than is predicted by the equations formulated by Galileo. This time, the axis of spin is horizontal and perpendicular to the motion of the ball. When the ball is driven off the tee, the head of the club spins the ball so that the bottom of the ball is spinning in the direction of flight. That reduces the airflow speed and, by the Bernoulli effect, increases the pressure below the ball, which creates an upward force on the ball that counteracts gravity. In fact, the ball is almost weightless as it flies through the air, as if the spin is giving it a helping hand carrying it down the fairway.

There's one extra ingredient that we haven't included, and it explains why Carlos's free kick swung to the left so late: the drag on the ball. As with the ups and downs of the lemming populations, the secret to Carlos's magic turns out to be a transition from chaotic to regular behavior. The airflow behind a soccer ball can be either chaotic or regular. The chaotic airflow is called turbulence and happens when the ball is travelling very fast. The regular airflow is called laminar flow and happens at slower speeds. Where the switch from one to the other kicks in depends on the type of ball.

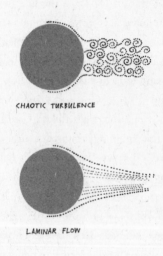

Figure 5.13   Chaotic turbulence causes less drag than regular "laminar" flow.

You can see the different sorts of airflow caused by different wind speeds quite easily. Walk in a straight line holding a flag (or a strip of fabric) so that it trails behind you, and watch it float along. Now do the same thing at a much faster speed, either by holding the flag out of a car window or by running as fast as you can into a strong wind. It will now be flapping around wildly. The reason is that air going around an object such as a flag behaves differently at different speeds. At lower speeds, the airflow is easily predictable, but at higher speeds, it's much more chaotic.

What effect does this change from turbulence to laminar flow have on taking a free kick? It turns out that chaotic turbulence causes much less drag on the ball. So when the ball is moving quickly, the spin doesn't have such a great effect on its direction, and the force of the spin is spread out over a larger part of the trajectory. When the ball slows and passes the transition point, the turbulence gives way to laminar flow, which causes much more drag. It's like someone slamming on the brakes. In that moment of transition, the air resistance increases by 150 percent. Now the effect of the spin can kick in, and the ball suddenly swerves much more dramatically. The extra drag also increases the lift, causing the Magnus effect to increase and pushing the ball even harder to the side.

Roberto Carlos needed a free kick far enough from the goal so that he could hit it hard enough to get the chaotic turbulence and for there to be time for the ball to slow down and bend before going out of play. When the ball is kicked at about 110 kilometers an hour, the airflow around it is chaotic, but about halfway through its flight, as it slows down, the turbulence changes. The brakes are applied, the ball's spin takes over, and Barthez is beaten.

It's not just games of soccer that are affected by this math. The way we travel is affected by chaos, too, particularly in the air. Most people associate the word *turbulence* with a request to fasten their seatbelt and being tossed to and fro by chaotic air currents. Airplanes travel much faster than a soccer ball, and the chaotic airflow over their wings—turbulent flow—increases the air resistance to the plane's flight, meaning more fuel has to be burned, at a higher cost.

One study concluded that a 10 percent reduction in turbulent drag could increase an airline's profit margin by 40 percent. Aeronautical engineers are always looking at ways to change the texture of a wing's surface to make the airflow less chaotic. One idea is to introduce a row of tiny parallel grooves along the wing, spaced as closely as the grooves on a vinyl record. Another is to cover the wing surface with minuscule tooth-like structures called denticles. Interestingly, the skin of a shark is covered with natural denticles, showing that nature discovered how to overcome fluid resistance long before engineers did.

Although it's been studied intensely, the turbulence behind a soccer ball or an airplane's wing is still one of the big mysteries of mathematics.

There is some good news: we've managed to write down the equations that describe the behavior of air or fluid. The bad news is that nobody knows how to solve them! These equations aren't important just to the likes of Beckham and Carlos. Weather forecasters need to solve them to predict air currents in the atmosphere, medics need to solve them to understand blood flow through the body, and astrophysicists need to solve them to figure out how stars move around in galaxies. All these things are controlled by the same mathematics. At the moment, forecasters, designers, and others can only use approximations, but because there is chaos hiding behind these equations, a small error can have a big effect on the outcome—so their predictions could be completely wrong.

These equations are called the Navier-Stokes equations, after the two nineteenth-century mathematicians who formulated them. They are not simple. A common representation of them is the following:

$$\frac{\partial}{\partial t} u_i + \sum_{j=1}^{n} u_j \frac{\partial u_i}{\partial x_j} = v \Delta u_i - \frac{\partial p}{\partial x_i} + f_i(x,t)$$

$$\operatorname{div} u = \sum_{i=1}^{n} \frac{\partial u_i}{\partial x_i} = 0$$

If you don't understand some of the symbols in these equations, don't worry—not many people do! But for those who know the language of math, these equations hold the key to predicting the future. They are so important that there is a million-dollar prize on offer for the first person to solve them.

The great German physicist Werner Heisenberg, who created quantum physics, once said, "When I meet God, I am going to ask him two questions: Why relativity? And why turbulence? I really believe he will have an answer for the first."

When Roberto Carlos was asked how he'd discovered the secret to bending balls so dramatically, he replied, "I have practised on my free-kick accuracy since I was a child. I used to stay at least one hour after each training session and practise extra on my free-kick accuracy. It's like with everything: the more pain and sweat, the more you will get."

I guess that applies to math as well. The more difficult the problem, the more satisfying it is when you crack it. So if the mathematical going gets tough, just remember the words of Roberto Carlos: "the more pain and sweat, the more you will get." And when you finally crack one of the big mathematical enigmas of all time, everyone will be thinking just what Barthez was thinking as he stared at the ball in the back of his net: how on Earth did you do that!

PICTURE CREDITS

2.36 Bagel © Raymond Turvey
2.37 Map of Europe
2.38 Seven shades © Joe McLaren
2.39 Unlocking rings © Raymond Turvey

CHAPTER THREE

3.1   Lizards © Joe McLaren
3.2   Lottery Ticket © Raymond Turvey
3.3   Winning Lottery Ticket © Raymond Turvey
3.4   Tetrahedral Dice © Raymond Turvey
3.5   Platonic Dice © Raymond Turvey
3.6   Icosahedron © Raymond Turvey
3.7   Archimedean Solid © Raymond Turvey
3.8   Dice pyramid © Raymond Turvey
3.9   Chocolate Roulette © Joe McLaren
3.10 Arranged chocolate Roulette © Joe McLaren
3.11 Cake stand © Raymond Turvey
3.12 Dürer's Magic Square © Joe McLaren
3.13 Bridge connections © Raymond Turvey
3.14 Envelope © Raymond Turvey
3.15 Bridges of Königsberg, eighteenth century © Joe McLaren
3.16 Bridges of Kaliningrad, twenty-first century © Joe McLaren
3.17 Traveling salesman problem © Raymond Turvey
3.18 Dinner party problem © Raymond Turvey
3.19 Country borders © Raymond Turvey
3.20 Minefields © Raymond Turvey
3.21 Minefields © Raymond Turvey
3.22 Loading truck problem © Joe McLaren
3.23 Traveling salesman solution © Raymond Turvey

CHAPTER FOUR

4.1   Babington Code © Joe McLaren
4.2   Enigma Machine © Joe McLaren
4.3   Chappe Machine © Joe McLaren
4.4   Chappe Code © Joe McLaren
4.5   Nelson semaphore © Raymond Turvey
4.6   Semaphore code © Joe McLaren
4.7   Beatles cover © Joe McLaren
4.8   Beatles cover corrected © Joe McLaren
4.9   CND Symbol
4.10 Morse Code Alphabet © Raymond Turvey
4.11 Morse Code © Raymond Turvey

CHAPTER FIVE

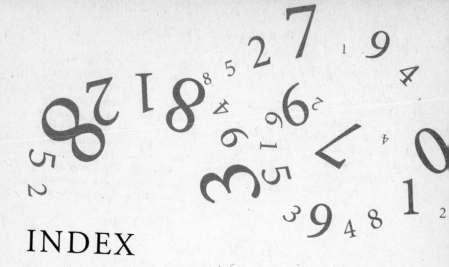

INDEX